Communications and Control Engineering

Series Editors

A. Isidori • J.H. van Schuppen • E.D. Sontag • M. Thoma • M. Krstic

For further volumes:
www.springer.com/series/61

Luminita Manuela Bujorianu

Stochastic Reachability Analysis of Hybrid Systems

 Springer

Luminita Manuela Bujorianu
School of Mathematics
University of Manchester
Manchester
UK

ISSN 0178-5354   Communications and Control Engineering
ISBN 978-1-4471-2794-9   e-ISBN 978-1-4471-2795-6
DOI 10.1007/978-1-4471-2795-6
Springer London Dordrecht Heidelberg New York

British Library Cataloguing in Publication Data
A catalogue record for this book is available from the British Library

Library of Congress Control Number: 2012932743

Mathematics Subject Classification (2010): 60J27, 60J25, 60J35, 60J40, 60J60, 60K15, 62F15, 60G40, 32U20, 93C30, 34A38, 49J40, 49J55, 49L20, 49L25, 35J99, 47B34, 45H05

Printed on acid-free paper

Springer is part of Springer Science+Business Media (www.springer.com)

In memory of Elena and Constanta

Preface

The concept of hybrid systems is a mathematical model for real life systems whose behaviour involves a mixture of discrete and continuous dynamics. Such systems are everywhere around us, from the simple bouncing ball to sophisticated cars, trains, planes and robots. Basically, every form of life exhibits a hybrid dynamics. Most physical and meteorological phenomena exhibit a hybrid dynamics. For scientists and engineers, the systems of interest are those for which the discrete and continuous dynamics interact or are connected. This interaction can take many forms, but the most common one occurs when discrete/digital controllers switch between different continuous processes. Other forms of interaction include discrete transitions that depend on continuous evolutions, or other appear as result of a decision process, or because of the occurrence of certain events. These systems are not given a distinct name. Most recently, the term cyber-physical systems has been proposed to denote collections of such interacting systems as networked or multi-agent hybrid systems. Although many engineered systems exhibit a mixture of discrete and continuous dynamics, in many cases the underlying mathematical model is not a hybrid system. We have to point out here that hybrid systems denote a family of mathematical models, but real life systems are traditionally developed using either continuous-based mathematical models (as in control engineering), or discrete-based mathematical models (as in computer science). Methods based on hybrid systems can be applied straightforwardly to practical systems. This is why system engineering that uses hybrid discrete continuous models (i.e. hybrid systems) has been subject to intensive research in the past two decades.

Research in hybrid systems is usually goal oriented. For example, in life sciences, hybrid systems are used mainly for modelling and analysis. In control engineering, the main emphasis is on design, optimisation, analysis and control synthesis. In computer science, the major focus is on modelling and formal verification. Consequently, three major specialisations of the hybrid system concept arose: hybrid dynamical systems, hybrid control systems and hybrid automata.

Traditionally, hybrid systems have been proposed as a modelling paradigm for embedded systems. Embedded systems represent an engineering paradigm for electronic systems operating in physical, often harsh environments. The simplest way to obtain a hybrid model for an embedded system is to model the electric system as a discrete automaton and the environment as a physical process governed by some differential equations. In many cases, the electronic system is a hybrid system itself and then the embedded system is modelled as a hybrid system operating in a physical environment. This is the case for most automotive systems. For example, a car or a plane has many electronic controllers that can switch between different modes of operation of a physical device according to the environment evolution.

Modern applications of hybrid systems have become increasingly complex. The complexity is due to rich interactions, complicated dynamics, randomness of environment, uncertainty of measurements and tolerance to faults. Traditionally, the way one reduces system complexity is by employing stochastics. Randomisation has become a standard method in modelling and analysis of complex systems. The standard theory of stochastic processes has three major avenues: the abstract case, which is then specialised to the discrete case, and the continuous case. One could easily remark that the hybrid discrete continuous case is missing in mathematical or engineering oriented text books. In fact, hybrid stochastics started developing only in the past decade.

The purpose of this book is to present recent developments in the use of hybrid discrete continuous stochastic modelling for the analysis of embedded systems. The book is based on an interdisciplinary perspective that can be used in control engineering and computer and life sciences. The modelling aspect is now known as the theory of stochastic hybrid systems. The analysis aspect is known as stochastic reachability. Stochastic hybrid systems constitute a family of models that resulted from various types of randomisation of hybrid systems. There are many ways to introduce randomness in a model. There are probabilistic hybrid systems with probability distributions associated only with discrete transitions. Other stochastic models consider the noise that perturbs the continuous evolutions. In the most general form, a stochastic hybrid system considers probability distributions for both discrete and continuous transitions, and moreover these distributions can depend on each other. The simplest form of stochastic hybrid systems was introduced in control engineering three decades ago, and in mathematics even earlier. The most general form of stochastic hybrid systems was introduced in the last decade, motivated by problems in air traffic control. Indeed, a turbulent environment can produce a significant deviation of a plane trajectory from its designated flight plan. In order to prevent serious problems like sector intrusions or collisions, corrective actions have to be taken. Such situations are described with high accuracy by stochastic hybrid systems.

The analysis of stochastic hybrid systems is very important since it is used in all activities related to system engineering. A specific type of analysis is developed in this context. It is called stochastic reachability analysis and it admits a neat mathematical formulation: Evaluate the probability that the system starting from a given initial state can reach a target state set.

In the case of discrete systems, this mathematical formulation describes the well established concept of transient analysis of Markov chains (or probabilistic model checking). In the continuous case, this problem is related to the classical theory of hitting probabilities, or first passage times. In the hybrid case, the system exhibits behaviours that violate all the properties that are known in discrete and continuous cases. Consequently, none of the classical methods can be directly applicable to hybrid systems.

The research goal in stochastic reachability analysis is twofold: to invent new specific methods or to develop ways to apply the existing methods from the discrete and continuous cases. This body of research is the subject of this book. Currently, no book is dedicated to stochastic reachability for hybrid systems. This book is ideal for postgraduate students, offering complete background, the state of the art in research, case studies and many open problems. Practitioners (engineers) will find this book useful as they attempt to construct more reliable flexible and robust systems. Although the book is intended as a solid theoretical presentation, there are many indications on how one may obtain tool support. The topic treated in this book is a highly interdisciplinary area, so knowledge of multiple disciplines is required. The book is self-contained and it addresses postgraduate students, researchers and practitioners from control engineering, computer science, and applied mathematics.

I would like to thank many people who made this book possible. First, I express my gratitude to Springer editors, Oliver Jackson and Kathy McKenzie. Oliver spotted the need for a monograph to introduce stochastic reachability analysis to a wider audience, and he challenged me to write such a book. His patience and constant encouragement have been invaluable for its completion. Kathy has so patiently studied the manuscript as to suggest dozen of improvements. Most importantly, I would like to thank Professor John Lygeros for introducing me to the area of stochastic hybrid systems. I am indebted to Savi Maharaj, Henk Blom, Rom Langerak, Holger Hermanns, Jianghai Hu, Xenofon Koutsoukos, Maria Prandini, Alesandro Giua and Joost-Pieter Katoen for fruitful collaboration and discussion. I also thank all colleagues from the Hybridge project for their role in introducing the stochastic reachability analysis concept. Professor Dave Broomhead made me evaluate this research in a multidisciplinary context, with support provided by the CICADA project at the University of Manchester. Thank you, Dave, for three years at Manchester and contagious enthusiasm!

Last, but not least, I thank my family for their immense support!

Manchester, UK   Luminita Manuela Bujorianu

Contents

Acronyms

ATC	Air traffic control
ATM	Air traffic management
CD	Conflict detection
CR	Conflict resolution
CTMC	Continuous time Markov chain
DTMC	Discrete time Markov chain
GSHS	General stochastic hybrid system
GSDP	General switching diffusion process
ODE	Ordinary differential equation
OSP	Optimal stopping problem
PDMP	Piecewise deterministic Markov process
PDE	Partial differential equation
PHA	Probabilistic hybrid automata
SDE	Stochastic differential equation
SDP	Switching diffusion process
SHA	Stochastic hybrid automata
SHS	Stochastic hybrid systems

List of Figures

Chapter 1
Introduction

'Noise had a glorious birth. It was one of the three miracles of the miracle year, 1905. Einstein, always aiming to solve the greatest of problems and to solve them simply, saw that noise could be the instrument to establish one of the greatest ideas of all time—the existence of atoms. In a few simple pages he invented noise, and thus "noise" was born. Immediately after Einstein, there was an incredible flurry of ideas of the most profound kind which continues to this day. Noise permeates every field of science and technology and has been instrumental in solving great problems, including the origin of the universe. But noise, considered by many as unwanted and mistakenly defined as such by some, has little respectability. The term itself conjures up images of rejection. Yet it is an idea that has served mankind in the most profound ways. It would be a dull, gray world without noise.'

Leon Cohen: The history of noise (On the 100th anniversary of its birth). *IEEE Signal Processing Magazine*, Vol. 20, November 2005.

Stochastic hybrid systems (SHS) can exhibit very complex behaviours, a fact that makes their analysis both critical and challenging. Verification of SHS is an area where deductive formal methods, relying on mathematical inferences and proofs to produce precise statements about a system, are indispensable. Formal methods are also needed in system synthesis, particularly when correctness, robustness and optimality are of paramount importance, which renders design by informal reasoning combined with trial and error ineffective.

Besides the more traditional properties such as stability and performance, properties of interest in (probabilistic or stochastic) hybrid systems also include safety and reachability. In principle, safety verification aims to show that starting from any initial condition in some prescribed set, a system cannot evolve to some unsafe region in the state space. On the other hand, reachability verification aims to show that for some initial conditions in some prescribed set, the system will evolve to some target region in the state space. The above properties are most relevant when system specifications are given in temporal logic formulas [131, 177]. These verification questions are by no means easy to answer; for very simple classes of hybrid systems they are known to be undecidable already [116].

L.M. Bujorianu, *Stochastic Reachability Analysis of Hybrid Systems*,
Communications and Control Engineering,
DOI 10.1007/978-1-4471-2795-6_1, © Springer-Verlag London Limited 2012

One of the most important goals of this monograph is to develop methods for verification of stochastic hybrid systems. Since these systems are specified through the use of complex tools of stochastic analysis, the first main step is to define the mathematical formulation of the verification problem in this context. Then we have to investigate mathematical properties, probabilistic and analytic solutions of this problem. The ultimate goal is to investigate if the formal verification methods can be further developed for SHS. This can be a long research track because SHS, being multidimensional systems (continuous time/state space, stochastic dimension, hybrid dimension), do not always admit finite state abstractions.

Chapter 1 is an introduction to this monograph. We need also to provide here a motivation. Why does one need to go through this diversity of methods and tools to get acquittance in the subject of stochastic reachability?

The book treats stochastic reachability problems for hybrid systems. Therefore, we have to keep in mind two main facets:

1. the models we work with;
2. the reachability issues in the probabilistic framework.

Regarding the first one, much material of this book is dedicated to the introduction of hybrid models with randomness features. Roughly speaking, hybrid system models 'decorated' with randomness are stochastic processes with jumps (discontinuous dynamics). Moreover, in order to study stochastic control problems (like the stochastic reachability problem) for these processes, we need the Markovian framework. The memoryless property (Markov property) of such processes is essential to consider when we mathematically formalise the control problems. Therefore, it is very important to find the right assumptions that ensure the Markovian nature of the random hybrid processes. To understanding these issues we dedicated three chapters. One is a background chapter on Markov processes theory (Chap. 2). The first objective of this chapter is to present two categories of Markov models: Markov models with discrete state space (usually called Markov chains), and Markov models with continuous state space (called Markov processes). For Markov processes we go further with the classification, and according to trajectory continuity criteria we may have continuous processes (with continuous trajectories) and processes with jumps (whose trajectories are piecewise continuous). The classical examples of continuous processes are the diffusion processes. In the class of processes with jumps we may include the so-called Markov jump processes (which are piecewise constant Markov processes) and diffusions with jumps. In this chapter, we give the main tools and results for Markov processes that are used further in this book. We have to point out here that the theory of Markov processes is rich and extensively large. We summarise only the results that will be used to model stochastic hybrid systems and to set up the stochastic reachability problem. In principle, all the definitions revolve around the concept of infinitesimal generator of the process, which is the analog of the stochastic matrix of a Markov chain.

Chapter 3 is about hybrid system modelling. We start with the classical deterministic hybrid systems and present two modelling perspectives: the dynamical system modelling approach (for people working in control engineering and applied mathematics) and the automata modelling approach (for people working in computer science). Then we show how and where randomness has to be considered. The last part of this chapter is dedicated to a classification of stochastic hybrid system models that have been developed in the literature. We need to consider again two perspectives. The dynamical system approach becomes now the stochastic process perspective. We have to present, in fact, stochastic processes with switchings and jumps. The automata approach has to define stochastic hybrid systems as hybrid automata enhanced with probability distributions. The two approaches are, in fact, two faces of the same coin. People working with hybrid automata develop the semantics of these automata using concepts from theories of dynamical systems or stochastic processes. People working in applied mathematics or control engineering start directly with the definitions of (stochastic) hybrid dynamical processes. Chapter 4 provides a complete study of the most general model of stochastic hybrid systems. All other models of stochastic hybrid systems presented in the previous chapter are special cases of this general model.

Chapter 5 is a necessary step on our way to defining stochastic reachability. This chapter presents different mathematical objects that can be linked with the characterisation of stochastic reachability of hybrid systems. Passing from deterministic to stochastic involves finding novel methodologies based on probability distribution handling. Of course, these probability distributions satisfy some 'deterministic equations', but these are partial differential equations, and it is well known that computational approaches for such equations are quite difficult. This chapter is at the heart of the book. We define formally the concept of reachability in the stochastic context. Mainly, the stochastic reachability problem is defined as a problem of estimating a measure of those process trajectories, which satisfy a 'visiting' condition of a given target set. It is very important to show that the definitions we give are consistent, i.e., the stochastic reachability problem is well posed. In this chapter, this reachability measure is mathematically characterised. Moreover, we give alternative ways to define the stochastic reachability measure. For example, one can be interested in measuring the sojourn time in a given set, or in finding the expectations or other statistical parameters of the hitting times corresponding to the target sets. There is also need to set up a kind of 'stochastic reachability dictionary'. This is because, there may be researchers working in different areas of control, stochastic analysis or computer science who use a different terminology for the same thing.

The next three chapters treat issues that represent different approaches on stochastic reachability analysis from three perspectives: probabilistic, analytic and statistical.

Chapter 6 is an overview of the most popular probabilistic methodologies existing for stochastic reachability analysis. This chapter is a gentle introduction to such approaches on stochastic reachability. The reader willing to play only with probabilistic methods can read just this chapter and then go to the appropriate reference to study in more detail the desired methodology. This chapter presents some approaches on stochastic reachability based only on stochastic analysis concepts associated to Markov processes. This is a very useful study since it gives a very good understanding of the problem and the taste of the mathematics behind it. At the end, we get a clear conclusion, namely if we restrict ourselves to work only on Markov processes ground, we obtain only upper/lower bounds of the reach probabilities. Better estimations of the reach probabilities can be obtained when we add to this framework elements of stochastic control, like the characterisations based on optimal control.

Chapter 7 develops further the existing analytic methods and then the computational methods that are based on dynamic programming. This chapter is the bridge between analytic solutions and computational ones. The essence of this connection is the characterisation of reach probabilities as solutions to some appropriate Hamilton–Jacobi–Bellman equations. Then, dynamic programming methods can be fruitfully used to estimate the reach set probabilities.

Chapter 8 presents a novel approach on stochastic reachability built in a Bayesian framework.

Markov chain approximation methods will be touched upon only in the application context, namely in air traffic control problems. The reason for this is that Markov chain approximations are mostly based on Kushner-type approximations [158]. To give a rigorous presentation of these approximation methods, we need more background, but then the intended reader might be confused. The research in finding the suitable Markov chain approximations for stochastic hybrid systems is still under development. Since these approximations often may suffer from the large number of states (state space explosion), approximation gain could be insignificant and, in fact, we end up replacing one difficult problem with another. The new approach in control theory regarding these problems is not to find discrete approximations but continuous ones that are 'smoother'. The simulation approaches on stochastic reachability are also mentioned in Chap. 5, but again these will not be developed further in the book since, in order to comprehend them, a reader would need background on Monte Carlo simulations.

Chapter 9 provides a new perspective on the issue treated in this book. Inspired by computer science approaches on Markov chains, we develop methods to 'simulate' stochastic hybrid systems by simpler models (called abstractions) such that the computation of reachable event probability is simplified by using these abstractions.

The last two chapters of the book present applications of stochastic reachability. Chapter 10 refines the concept of stochastic reachability by adding constraints regarding space and time. The last chapter presents more relevant applications in aerospace engineering, with explanations of the state-of-the-art research in this area. This chapter is intended as a summary of research that promotes the use of stochastic hybrid systems and stochastic reachability in air traffic control and biology.

Chapter 2
Markov Models

2.1 Overview

The theory of Markov processes can be divided into subtheories that depend on whether time is discrete or continuous or whether the set of possible states is finite or countably infinite. In this case, we speak about *Markov chains*. If the set of possible states has cardinality of the continuum, we speak of *continuous Markov processes*.

2.2 Discrete Space Markov Models

A *Markov chain* is a stochastic process with the property that any subsequent state depends only on the state preceding it. It is a particular type of Markov process, in which the process has a discrete (finite or countable) number of states.

A Markov chain can have a discrete set of times, and in this case is called a discrete time Markov chain (DTMC); or it can have a continuous set of times, and so is called a continuous time Markov chain (CTMC).

2.2.1 Discrete Time Markov Chains

Formally, let us consider processes with a countable or finite state space $\mathbb{S} \subset \{0, 1, 2, \ldots\}$.

A discrete time discrete, state stochastic process $\{X_n : n \geq 0\}$ with a state space $\mathbb{S}$ is called a *discrete time Markov chain* if and only if it has a Markov property,

$$\mathbb{P}(X_{n+1} = j | X_n = i; X_{n-1} = i_{n-1}, \ldots, X_0 = i_0) = \mathbb{P}(X_{n+1} = j | X_n = i) \quad (2.1)$$

for all $n \geq 0$; $i, j, i_0, i_1, \ldots, i_{n-1} \in \mathbb{S}$ (i.e., the probability of the next state, given the current state and the entire past, depends only on the current state).

The Markov property (2.1) states that, given the evolution of the process $\{X_n : n \geq 0\}$ up to any 'current' time n, the probabilistic description of its behaviour at

L.M. Bujorianu, *Stochastic Reachability Analysis of Hybrid Systems*,
Communications and Control Engineering,
DOI 10.1007/978-1-4471-2795-6_2, © Springer-Verlag London Limited 2012

time $n + 1$ (and, by induction, the probabilistic description of all subsequent behaviour) depends only on the current state $X_n = i$ and not on the previous history of the process.

We say that a Markov chain $\{X_n : n \geq 0\}$ is *time-homogeneous* if the conditional probabilities (2.1) are independent of n for all $i, j \in \mathbb{S}$; that is, for all $i, j \in \mathbb{S}$ there exist p_{ij} such that

$$\mathbb{P}(X_{n+1} = j | X_n = i) = p_{ij}$$

for all $n \geq 0$.

The numbers p_{ij} are called *one-step transition probabilities* and the matrix $P = (p_{ij})_{i,j\in\mathbb{S}}$ is called a *one-step transition matrix*, or simply the *stochastic matrix* of the Markov chain $\{X_n\}$. Necessarily, we have

$$p_{ij} > 0 \quad \text{and} \quad \sum_{j\in\mathbb{S}} p_{ij} = 1$$

for all $i \in \mathbb{S}$. The number p_{ij} is interpreted as the probability that the state of the process changes (has a 'transition') from state i to state j in one time step.

The initial state probabilities

$$p_i = \mathbb{P}(X_0 = i) \tag{2.2}$$

for all $i \in \mathbb{S}$ can be organised in an initial state vector

$$p = (p_i)_{i\in\mathbb{S}}$$

called *initial probability distribution* of the given Markov chain.

The sample paths of a Markov chain are completely characterised by the one-step transition probabilities, p_{ij}, $i, j \in \mathbb{S}$ and initial-state probabilities, p_i, $i \in \mathbb{S}$. Given values for these probabilities, the joint probability of any finite sample path can be decomposed into a product of an initial state probability and the appropriate sequence of transition probabilities.

Marginal Distributions   To predict the state of the system at time n; $n \geq 0$; we need to compute the *marginal distribution* of X_n:

$$p_j^{(n)} = \mathbb{P}(X_n = j) = \sum_{i\in\mathbb{S}} \mathbb{P}(X_n = j | X_0 = i)\mathbb{P}(X_0 = i)$$

$$= \sum_{i\in\mathbb{S}} \mathbb{P}(X_n = j | X_0 = i) p_i$$

where p_i are the initial state probabilities given by (2.2). Therefore, the marginal distributions can be computed if the following conditional probabilities, called *n-step transition probabilities*, can be calculated:

$$p_{ij}^{(n)} = \mathbb{P}(X_n = j | X_0 = i)$$

for $i, j \in \mathbb{S}$. For the n-step transition probabilities, we use the matrix notation $P^{(n)} = (p_{ij}^{(n)})_{i,j\in\mathbb{S}}$. Also, the marginal distributions can be collected in a vector $p^{(n)} = (p_j^{(n)})_{j\in\mathbb{S}}$.

Chapman–Kolmogorov Equations   Based on the Markov property, it can be proved that the n-step transition probabilities satisfy the so-called *Chapman–Kolmogorov equations*, i.e.,

$$p_{ij}^{(n)} = \sum_{r \in \mathbb{S}} p_{ir}^{(k)} p_{rj}^{(n-k)}$$

for $i, j \in \mathbb{S}$, where k is a fixed integer such that $0 \leq k \leq n$. Using the Chapman–Kolmogorov equations, the following equalities can be proved by induction:

$$P^{(n)} = P^n, \quad n \geq 0,$$
$$p^{(n)} = p P^n, \quad n \geq 0$$

where P^n is the nth power of P.

2.2.2 Continuous Time Markov Chains

Continuous time and finite state space Markov chains, well known for years, represent a model of perfectly regular random evolution, which stays in a state for a certain period of time (of known law), then jumps into another state drawn at random according to a known law, and so on indefinitely.

CTMCs are stochastic processes that behave in the following way. Suppose that at time $t \geq 0$, the system enters state i. Then it stays in state i for a random amount of time called the *sojourn time* and then jumps to a new state $j \neq i$ with probability p_{ij}. The Markovian property is illustrated by the fact that the sojourn time and the new state depend only on i and not on the history of the system prior to time t. Furthermore, given that the current state is i; the sojourn time and the new state are independent of each other.

Just as with discrete time, a continuous time stochastic process with a discrete state space, which we can take to be the positive integers, is a CTMC if the conditional probability of a future event given the present state, and additional information about past states depends only on the present state.

Formally, a stochastic process $\{X(t) : t \geq 0\}$ (with an integer state space) is a CTMC if

$$\mathbb{P}\{X(t+s) = j \,|\, X(s) = i, X(r) = i_r, r < s\} = \mathbb{P}\{X(t+s) = j \,|\, X(s) = i\}$$

for all states i and j and for all times $t \geq 0$ and $s \geq 0$.

The conditional probábilities $\mathbb{P}\{X(t+s) = j | X(s) = i\}$ are called the *transition probabilities*. We will consider the special case of *stationary transition probabilities* (sometimes referred to as *homogeneous transition probabilities*), occurring when

$$\mathbb{P}\{X(t+s) = j \,|\, X(s) = i\} = \mathbb{P}\{X(t) = j \,|\, X(0) = i\} = p_{ij}(t)$$

for all states i and j, and for all times $t \geq 0$ and $s \geq 0$.

The concept of transition probabilities is essential in the analysis of CTMC. However, the computation of the transition probabilities is not straightforward and other analytical tools might be necessary.

In a similar way, as in discrete time, the evolution of the transition probabilities over time is described by the Chapman–Kolmogorov equations, but they take a different form in continuous time.

Chapman–Kolmogorov Equations   For all times $t \geq 0$ and $s \geq 0$, we have

$$p_{ij}(t + s) = \sum_k p_{ik}(t) p_{kj}(s).$$

Using matrix notation, we write $P(t)$ for the square matrix of transition probabilities $(p_{ij}(t))$ and call it the *transition function*. In matrix notation, the Chapman–Kolmogorov equations reduce to a simple relation describing the semigroup property of the transition function involving the following matrix multiplication:

$$P(t + s) = P(t)P(s)$$

for all times $t \geq 0$ and $s \geq 0$.

A CTMC is well-specified when the following items are given:

(1) its initial probability distribution

$$p_i = \mathbb{P}(X(0) = i)$$

for all states i; and

(2) its transition probabilities $p_{ij}(t)$ for all states i and j and positive times t. We can use these two elements to compute the distribution of $X(t)$ for each t, namely,

$$p_j(t) = \mathbb{P}(X(t) = j) = \sum_i p_i p_{ij}(t).$$

In continuous time, it is possible for chains to behave strangely, e.g., to run through an infinite number of states in finite time. We exclude such possibilities and consider only CTMC for which the transition probabilities, $p_{ij}(t)$, are differentiable at $t = 0$. Such chains can be specified by a matrix Q, called the *generator* of the chain, which has elements q_{ij} satisfying

$$p_{ij}(t) = \delta_{ij} + q_{ij}t + o(t),$$

where δ_{ij} is the Kronecker delta ($\delta_{ii} = 1$ and $\delta_{ij} = 0$ if $i \neq j$). Since $p_{ij}(t)$ is a probability, $q_{ij} \geq 0$; $q_{ii} \leq \cdot 0$ and $\sum_j q_{ij} = 0$, so that $q_{ii} = -\sum_{j \neq i} q_{ij}$. Note that the interpretation of q_{ij} is the probability per time unit that the system makes a transition from state i to state j, or, in other words, the transition rate or transition intensity of the process to jump from i to j. Recall that an expression $A(t)$ is $o(t)$ if $\lim_{t \to 0} \frac{A(t)}{t} = 0$. In terms of matrices, the dependence between the transition function and the generator can be expressed as follows:

$$P(t) = I + Qt + \mathbf{o}(t) \tag{2.3}$$

where the matrix $\mathbf{o}(t)$ is the matrix of $o(t)$ terms.

The relation (2.3) is equivalent to

$$q_{ij} = p'_{ij}(0+) = \frac{dp_{ij}(t)}{dt}\bigg|_{t=0+},$$

where $0+$ denotes the right derivative since $p_{ij(t)}$ is not defined for $t < 0$.

Therefore, we can specify a CTMC model via the transition rate matrix Q, in the same way as a DTMC is specified via the matrix P.

To construct the transition probabilities $p_{ij}(t)$ from the transition rates $q_{ij} = p'_{ij}(0+)$, we apply Chapman–Kolmogorov equations to show that the transition probabilities satisfy two systems of ordinary differential equations (ODEs) generated by the transition rates. In matrix notation, these will be simple first order linear ODEs.

Theorem 2.1 (Kolmogorov forward and backward ODEs) *The transition probabilities satisfy both*

- *the Kolmogorov forward differential equations*

$$p'_{ij}(t) = \sum_k p_{ik}(t)q_{kj} \quad \text{for all } i, j,$$

or, in matrix notation,

$$P'(t) = P(t)Q; \tag{2.4}$$

- *and the Kolmogorov backward differential equations*

$$p'_{ij}(t) = \sum_k q_{ik}p_{kj}(t) \quad \text{for all } i, j,$$

or, in matrix notation,

$$P'(t) = QP(t). \tag{2.5}$$

The matrix equations (2.4) and (2.5) have exponential matrix solutions. Subject to the boundary condition $P(0) = I$ (I is the identity matrix), these equations have the formal solution (care is needed when $\mathbb{S}$ is not finite)

$$P(t) = \exp(tQ) := \sum_{n=0}^{\infty} \frac{t^n Q^n}{n!},$$

where $Q^0 = I$. Therefore, the transition probabilities are specified by the matrix of transition rates Q, and so the chain is specified by Q and the initial distribution $p(0)$.

The Embedded DTMC   Given Q and an initial state i, the chain leaves state i at a rate $\sum_{j \neq i} q_{ij} = -q_{ii}$ and hence stays in i for a holding time that is exponentially distributed with mean $-1/q_{ii}$. When the chain leaves the state i, it jumps in a state $j \neq i$ with probability $(-\frac{q_{ij}}{q_{ii}})$. Then the sojourn time of the chain in the state j will be exponentially distributed with mean $-1/q_{jj}$, before jumping in a new state k with probability $(-\frac{q_{jk}}{q_{jj}})$, and so on.

This particular description of a CTMC is useful mostly for simulations. The sequence of visited states forms a DTMC (called the *embedded jump chain*) with transition probabilities

$$\widetilde{p}_{ij} = \begin{cases} -\frac{q_{ij}}{q_{ii}}, & \text{if } i \neq j, \\ 0, & \text{if } i = j. \end{cases}$$

Uniformisation   For all states i, let us use the following notation:

$$q_i := -q_{ii}.$$

The uniformisation method, usually used for the transient analysis of CTMCs, can be thought of as another way to describe a CTMC. By uniformisation, a CTMC is transformed into another CTMC where all the states have the same mean holding time $1/q$. This is done by allowing 'fictitious' transitions from a state to itself. This new CTMC is equivalent to a DTMC, after normalisation of the rows, together with an associated Poisson process of rate q. The one-step transition matrix $\widehat{P}$ of the DTMC is given by

$$\widehat{P} = \frac{Q}{q} + I, \tag{2.6}$$

where $q > \max_i q_i$. The number of transitions in the DTMC that occur in a given time interval is given by a Poisson process with rate q.

When the CTMC is in state i, each of these potential transitions is a real transition (to another state) with probability q_i/q whilst the potential transition is a fictitious transition (a transition from state i back to state i, meaning that we remain in state i at that time) with probability $1 - q_i/q$, independent of past events. In other words, in each state i, we perform independent thinning of the Poisson process having rate q, creating real transitions in state i according to a Poisson process having rate q_i, just as in the original model.

Uniformisation is useful because it allows us to apply properties of DTMCs to analyse CTMCs. Moreover, for the general CTMC characterised by the rate matrix Q, we have transition probabilities $p_{ij}(t)$ expressed via $\widehat{P}$ given by (2.6) and q as follows:

$$p_{ij}(t) = \sum_{n=0}^{\infty} \frac{e^{-qt}(qt)^n}{n!} \widehat{p}_{ij}^{(n)}, \quad t \geq 0,$$

where $\widehat{p}_{ij}^{(n)}$ is the corresponding entry of $\widehat{P}^{(n)}$, which is the n-step transition matrix of the uniformised Markov chain.

Important Examples   The most prominent examples of CTMCs are the Poisson Processes and the birth-and-death processes (see the definitions in the Appendix).

Let us consider a birth-and-death process with the birth rate $\{\lambda_i\}_{i=0,1,\ldots,\infty}$ and the death rate $\{\mu_i\}_{i=0,1,\ldots,\infty}$. The transition function of a birth-and-death process is defined as follows:

$$p_{ij}(h) = \begin{cases} \lambda_i h + o(h), & \text{if } j = i + 1 \\ \mu_i h + o(h), & \text{if } j = i - 1 \\ 1 - (\mu_i + \lambda_i)h + o(h), & \text{if } j = i \\ o(h), & \text{otherwise.} \end{cases}$$

Then the transition rates are

$$q_{ij} = \begin{cases} \lambda_i, & \text{if } j = i + 1 \\ \mu_i, & \text{if } j = i - 1 \\ -(\mu_i + \lambda_i), & \text{if } j = i \\ 0, & \text{otherwise} \end{cases}$$

and the stochastic matrix is

$$Q = \begin{pmatrix} -\lambda_0 & \lambda_0 & 0 & 0 & \cdots \\ \mu_1 & -(\lambda_1 + \mu_1) & \lambda_1 & 0 & \cdots \\ 0 & \mu_2 & -(\lambda_2 + \mu_2) & \lambda_2 & \cdots \\ 0 & 0 & \mu_3 & -(\lambda_3 + \mu_3) & \cdots \\ \vdots & \vdots & \vdots & \vdots & \end{pmatrix}.$$

2.3 Continuous Space Markov Models

Usually, the stochastic processes that appear in this book take values in open subsets of different Euclidean spaces. In this case, these spaces might have nice topological properties. Most of the time, we will work with Markov processes defined on Lusin spaces. A *Lusin space* is a topological space, which is homeomorphic[1] with a Borel subset of a compact metric space. In measure theory, a Borel set is any set in a topological space that can be formed from open sets (or, equivalently, from closed sets) through the operations of countable union, countable intersection, and relative complement. Borel sets are named after Émile Borel. For a topological space **X**, the collection of all Borel sets on **X** forms a σ-algebra, known as the Borel algebra or Borel σ-algebra. The Borel algebra on **X** is the smallest σ-algebra containing all open sets (or, equivalently, all closed sets).

When we study Markov processes we may need to deal with different topological spaces like:

- *Polish space:* A *Polish space* is a topological space homeomorphic with a complete, separable metric space.[2]
- *Borel space:* A *Borel space* is a topological space homeomorphic to a Borel subset of a complete separable metric space.

---

[1]In the mathematical field of topology, a homeomorphism or topological isomorphism is a continuous function between topological spaces that has a continuous inverse function.

[2]I.e., it has a countable dense subset.

- *Analytic space*: An *analytic space* is a topological space that is the image of a Polish space under a continuous function from one Polish space to another.

A topological space **X** is called a *Lusin* (respectively, *Souslin, Radon*) topological space if it is homeomorphic to a Borel (respectively, analytic, universally measurable) subset of a compact metric space. Note that since every Borel subset of a compact metric space is analytic and every analytic set is universally measurable, every Lusin/Souslin space is Radonian.

The classical theory of Markov processes is typically carried out in the setting of Polish spaces rather than on abstract measure spaces. The analytic spaces generalise Polish spaces.

In what follows, $(\Omega, \mathscr{F}, \mathbb{P})$ will denote a probability space, **X** a topological (Lusin/Polish/analytic) space, $\mathscr{B}$ its Borel σ-algebra and T an interval of the real line. For each $t \in T$,

$$x_t : (\Omega, \mathscr{F}, \mathbb{P}) \to (\mathbf{X}, \mathscr{B})$$

is a measurable function. Usually, $T := [0, \infty)$.

Let $\mathscr{B}^b(\mathbf{X})$ be the lattice of bounded positive measurable functions on **X**. This is a Banach space under the norm

$$\|f\| = \sup_{x \in \mathbf{X}} |f(x)|.$$

There are several different but essentially equivalent ways to describe continuous time Markov processes. We will start with the description based on transition probability functions.

Transition Functions   A function $p : \mathbf{X} \times \mathscr{B} \to [0, 1]$ is a *transition probability*, or a *stochastic kernel*, if

(i) $p(x, \cdot)$ is a probability measure in $\mathscr{B}$, and
(ii) $p(\cdot, B)$ is measurable for each $B \in \mathscr{B}$.

A *transition function* is a family $\{p_{s,t} \mid (s, t) \in T^2, s < t\}$ that satisfies, for each $s < t < u$, Chapman–Kolmogorov equation:

$$p_{s,u}(x, B) = \int p_{t,u}(y, B) p_{s,t}(x, dy).$$

A transition function is *time-homogeneous* if $p_{s,t} = p_{s',t'}$ whenever $t - s = t' - s'$. In this case we write p_{t-s} instead of $p_{s,t}$.

Let $\mathscr{M}_t \subset \mathscr{F}$ be an admissible filtration for a stochastic process M. M is Markov with the transition function $p_{s,t}$ if, for each nonnegative Borel measurable $f : \mathbf{X} \to \mathbb{R}$ and each $(s, t) \in T^2$, $s < t$,

$$\mathbb{E}[f(x_t)|\mathscr{M}_s] = \int f(y) p_{s,t}(x_s, dy).$$

Given a transition function $p_{s,t}$ on $(\mathbf{X}, \mathscr{B})$ and a probability measure μ on $(\mathbf{X}, \mathscr{B})$, there exists a unique probability measure $\mathbb{P}_\mu$ on $(\mathbf{X}^{[0,\infty)}, \mathscr{B}^{[0,\infty)})$, such that the coordinate process is Markov with respect to the σ-algebra $\sigma(x_u | u \leq t)$, with transition function $p_{s,t}$ and the distribution of x_0 given by μ [183].

Suppose we are given a homogeneous transition function p_t and an initial probability $\mu = \delta_x$, where δ_x denotes the Dirac measure at $x \in \mathbf{X}$ defined by

$$\delta_x(A) := \begin{cases} 1, & \text{if } x \in A \\ 0, & \text{if } x \notin A. \end{cases}$$

We can then construct a Markov process

$$M = \{x_t | t \in \mathbb{R}_+\}$$

having p_t as its transition function and

$$\mathbb{P}(x_0 = x) = 1.$$

It is, however, possible that the same construction $\mu = \delta_{x'}$ for some $x' \neq x$ would lead to a process defined on some other probability space.

A *Markov family* is a collection

$$\left(\Omega, \mathscr{F}, \mathscr{F}_t, x_t, (\mathbb{P}_x)_{x \in \mathbf{X}}\right),$$

where

(i) $(\Omega, \mathscr{F})$ is a measurable space,
(ii) $(\mathscr{F}_t)$ a filtration,
(iii) $(x_t | t \in \mathbb{R}_+)$ a family of $\mathbf{X}$-valued random variables such that x_t is $\mathscr{F}_t$-measurable for each t, and
(iv) $\mathbb{P}_x$ (for each $x \in \mathbf{X}$) is a probability measure on $(\Omega, \mathscr{F})$ such that (x_t) is a Markov process on $(\Omega, \mathscr{F}, \mathbb{P}_x)$ with transition function p_t and initial distribution δ_x, i.e.,

$$\mathbb{P}_x(x_0 = x) = 1.$$

$\mathbb{P}_x$ is called *Wiener probability*. We write $\mathbb{E}_x$ for the expectation with respect to $\mathbb{P}_x$. Note that the measure $\mathbb{P}_x$ and the transition function p_t are related by

$$p_t(x, A) = \mathbb{P}_x(x_t \in A). \tag{2.7}$$

In a Markov family, only the measure $\mathbb{P}_x$ depends on the initial point $x \in \mathbf{X}$; all the other ingredients are the same for every x. This provides yet another way of expressing the Markov property: Because the transition function is the same for every $\mathbb{P}_x$ we easily see that for $f \in \mathscr{B}^b(\mathbf{X})$ and $s, t \geq 0$, we have

$$\mathbb{E}_x\left[f(x_{t+s}) | \mathscr{F}_s\right] = \mathbb{E}_{x_s}\left[f(x_t)\right]$$

Thus, the behaviour of the process beyond time s is just another process started at x_s.

From now on we will generally consider Markov families rather than Markov processes, and use the notation

$$M = \left(\Omega, \mathscr{F}, \mathscr{F}_t, x_t, (\mathbb{P}_x)_{x \in \mathbf{X}}\right)$$

for a Markov process.

2.3.1 Strong Markov Processes

In this monograph, we make use of some standard notions in Markov process theory including underlying probability space, natural filtration, translation operator, Wiener probabilities, admissible filtration, stopping time and strong Markov property [33]. Most of these concepts are presented in the Appendix. In the following, we briefly present a version of the Markov property with respect to stopping times, called the *strong Markov property*.

Suppose that

$$M = (\Omega, \mathscr{F}, \mathscr{F}_t, x_t, \theta_t, \mathbb{P}_x)$$

is a Markov process. The state space of M is given by $(\mathbf{X}, \mathscr{B})$. Here, $(\Omega, \mathscr{F}, \mathbb{P}_x)$ denotes the sample probability space for each process with initial start point x. The family of σ-algebras $\{\mathscr{F}_t^0\}$ denotes the *natural filtration*, i.e., $\mathscr{F}_t^0 = \sigma\{x_s, s \leq t\}$. This is the smallest σ-algebra in $\mathscr{F}$ with respect to all the random variables $x_s, s \in [0, t]$ are measurable. Let us take

$$\mathscr{F}_\infty^0 = \bigvee_t \mathscr{F}_t^0.$$

The trajectories of M are modelled by a family of $\mathbf{X}$-valued random variables (x_t), which, as functions of time, have some continuity properties. This means that

(a) for each $t > 0$ the function $x_t : (\Omega, \mathscr{F}) \to (\mathbf{X}, \mathscr{B})$ is an $\mathscr{F}^0/\mathscr{B}$-measurable function and

(b) $t \to x_t(\omega)$ might be continuous, right continuous or right continuous with left limits (the càdlàg property).

Then $\mathscr{F}_t^0$ is the *minimum admissible filtration*.
The *shift operator*, or *translation operator*,

$$\theta_t : \Omega \to \Omega,$$

for all $t \geq 0$, has the property

$$x_s \circ \theta_t = x_{t+s}, \quad t, s \geq 0.$$

We adjoin an extra point Δ (the *cemetery point*) to $\mathbf{X}$ as an isolated point:

$$\mathbf{X}_\Delta = \mathbf{X} \cup \{\Delta\}.$$

The existence of Δ is assumed to have a probabilistic interpretation of $\mathbb{P}_x(x_t \in \mathbf{X}) < 1$, i.e., at some 'termination time' $\zeta(\omega)$ when the process M escapes to and is trapped at Δ.

If μ is a probability measure on $(\mathbf{X}, \mathscr{B})$, written as $\mu \in \mathscr{P}(\mathbf{X})$, then we can define

$$\mathbb{P}_\mu(\Lambda) = \int_{\mathbf{X}_\Delta} \mathbb{P}_x(\Lambda)\mu(dx), \quad \Lambda \in \mathscr{F}^0.$$

We then denote by $\mathscr{F}$ (respectively, $\mathscr{F}_t$) the completion of $\mathscr{F}_\infty^0$ (respectively, $\mathscr{F}_t^0$) with respect to all $(P_\mu)_{\mu \in \mathscr{P}(\mathbf{X})}$. This has the advantage that the class of null sets is the same for every $t \in \mathbb{R}_+$. From now on, we will refer to the family $\{\mathscr{F}_t\}_t$ as the *natural filtration* of M.

The *strong Markov property* means that the Markov property is still true with respect to the stopping times of the process M. For an admissible filtration $\{\mathscr{M}_t\}$, we say that M is *strong Markov* with respect to $\{\mathscr{M}_t\}$ if $\{\mathscr{M}_t\}$ is right continuous and

$$\mathbb{P}_\mu(x_{\tau+t} \in E | \mathscr{M}_\tau) = \mathbb{P}_{x_\tau}(x_t \in E), \tag{2.8}$$

where (2.8) holds P_μ almost sure (P_μ-a.s.), for all $\mu \in \mathscr{P}(\mathbf{X})$, $E \in \mathscr{B}$, $t \geq 0$ and for any $\{\mathscr{M}_t\}$-stopping time τ.

If the index set T is discrete, then the strong Markov property is implied by the ordinary Markov property. If time is continuous, this is not necessarily the case. It is generally true that if M is Markov and takes only countably many values, then M is strongly Markov. Therefore, any Markov chain is strongly Markov. However, in continuous time, the simple Markov property does not imply the strong Markov property. For examples of processes that are only Markov, but not strongly Markov, the interested reader may consult [208].

2.3.2 Continuous Processes

Diffusion Processes   A stochastic process $\mathbf{X}$ adapted to a filtration $\mathscr{F}$ is a diffusion when it is a strong Markov process with respect to $\mathscr{F}$, homogeneous in time and has continuous sample paths.

Diffusions are important because they are very natural models of many important systems: the motion of physical particles, fluid flows, noise in communication systems, financial time series, etc. Moreover, many discrete Markov models have large-scale limits which are diffusion processes. These are important in physics and chemistry, population genetics, queueing and network theory, certain aspects of learning theory, etc. These limits are often more tractable than more exact finite-size models.

One may postulate that the stochastic process driving a model has certain infinitesimal characteristics. Specifically, if the Markov process $\mathbf{X} = \{x_t | t \in [t_0, T]\}$ has the state space $\mathbb{R}^n$ and continuous sample paths with probability 1, it is called a *diffusion process* if its transition probability $p(s, x, t, B)$ is smooth, i.e., it satisfies the following three conditions for every $s \in [t_0, T]$, $x \in \mathbb{R}^n$ and $\varepsilon > 0$:

(i) $\lim_{t \searrow s} \frac{1}{t-s} \int_{|y-x|>\varepsilon} p(s, x, t, dy) = 0$

(ii) $\lim_{t \searrow s} \frac{1}{t-s} \int_{|y-x|\leq\varepsilon} (y - x) p(s, x, t, dy) = a(s, x)$

(iii) $\lim_{t \searrow s} \frac{1}{t-s} \int_{|y-x|\leq\varepsilon} (y - x)(y - x)^T p(s, x, t, dy) = B(s, x)$

where $a(s, x)$ and $B(s, x)$ represent well-defined $\mathbb{R}^n$- and $\mathbb{R}^{n \times n}$-valued functions, respectively. These functions, called the coefficients of the diffusion process, are referred as follows: a is called the (infinitesimal) *drift coefficient* and B the (infinitesimal) *diffusion* (or *covariance*) *coefficient* that should be positive definite.

The reason for this terminology is that processes of this kind were first encountered in physics in studying diffusion phenomena. Diffusion processes have the strong Markov property.

A Wiener process is also a special case of a strong diffusion process and, in fact, a particular kind of a continuous time strong Markov process.

Stochastic Differential Equations   The easiest way to obtain diffusions is through the theory of stochastic differential equations that are, roughly speaking, the result of adding a noise term to the right-hand side of a differential equation.

Ordinary differential equations which have the general form

$$\frac{dx}{dt} = f(t, x) \tag{2.9}$$

provide simple deterministic descriptions of the laws of motion of physical systems. The solution $x(t)$ of an initial value problem consisting of (2.9) together with the initial value

$$x(t_0) = x_0 \tag{2.10}$$

represents the state of such a system at time $t > t_0$, given that the state (2.10) was attained at time t_0. If random aspects in the physical system are to be considered, a number of modifications can be made in the formulation of the initial value problem (2.9), (2.10) as follows:

- initial point x_0 may be replaced by a random variable X_0;
- deterministic function $f(t, x)$ may be replaced by a random function $F(t, X, Y)$, where $Y = Y(t)$ designates a random input process uncoupled with the solution variable X;
- Y may represent the random coefficients of a linear/nonlinear operator whose form is specified by F.

The first of these three randomisation methods for (2.9) and (2.10) is exemplified by the motion of a space vehicle whose state consisting of position and momentum components changes according to a deterministic law, but whose initial values may be subject to some uncertainty. An AC electric power circuit with state described by voltages and phase angles of nodes whose rates of change are forced by noise is a particular case of the second type of randomness. An example of the third type arises when we consider intrinsic birth-death rates and interaction rates as stochastic processes in differential equation models of multispecies population evolution; a random initial value problem in which the stochastic input is coupled with the solution results.

The term *random differential equation* is regarding the last of these three types, whilst the *stochastic differential equation* concept considers equations of the second type, which are driven by noise and interpreted mathematically as *Itô equations*.

The ultimate scope of the analysis of any random initial value problem

$$\frac{dX(t)}{dt} = F\big(t, X(t), Y(t)\big), \tag{2.11}$$

$$X(t_0) = X_0 \tag{2.12}$$

generally, is to obtain the distribution of the solution process $X(t)$ in terms of the distributions of X_0, $Y(t)$, and the statistical/deterministic properties of F; determining the sample path structure of $X(t)$ is rather difficult. When randomness enters the problem only through the initial condition (a situation sometimes called 'cryptodeterministic') the solution is a deterministic transformation of the random variable X_0. The mean square theory for the random initial value problem, in this instance, is a direct analog of the ODE theory. For ODE, closed-form expressions for solutions are often difficult to get, and so we have to be happy with numerical approximations only, or less than complete qualitative information about solutions. Therefore, similar types of result should be expected for the random or stochastic differential equation case.

In general, the function F from (2.11) is given by

$$F\big(t, X(t), Y(t)\big) = b\big(t, X(t)\big) + \sigma\big(t, X(t)\big)Y(t), \tag{2.13}$$

with $Y(t)$ representing a Gaussian white noise process. The definition of the stochastic integral in the corresponding integral equation shows that such an equation is at best mathematically ambiguous. More precisely, usually the interest is to interpret (2.11) and (2.13) as the Itô equation [7]

$$dX(t) = b\big(t, X(t)\big) dt + \sigma\big(t, X(t)\big) dW(t), \tag{2.14}$$

where $W(t)$ denotes a Wiener or Brownian motion process; in (2.14) b and σ are deterministic functions, but with slight modifications, the theory extends to explicitly random functions. Stochastic differential equations (SDEs) were introduced by K. Itô in 1942 [136], and the basic theory was developed independently by Itô and I. Gihman during the 1940s [106]. Applications to control problems in electrical engineering motivated by the need for more sophisticated models spurred further work on these equations in the 1950s and 1960s. In the latter period, applications have been extended to many other areas including population dynamics in biology.

Briefly, the following definition presents formally the stochastic differential equation structure on the Euclidean space $\mathbb{R}^d$.

Let $b : \mathbb{R}^+ \times \mathbb{R}^d \to \mathbb{R}^d$ and $\sigma : \mathbb{R}^+ \times \mathbb{R}^d \to \mathbb{R}^{d \times m}$ be measurable functions (vector- and matrix-valued, respectively), (W_t) an m-dimensional Wiener process and $\mathbf{x}_0 \in \mathbb{R}^d$.

Definition 2.2 (SDE strong solution) An $\mathbb{R}^d$-valued stochastic process (x_t) on $\mathbb{R}^+$ is called a (strong) solution to the stochastic differential equation

$$\begin{cases} dx_t = b(t, x_t) + \sigma(t, x_t)\, dW_t \\ x_0 = \mathbf{x}_0 \end{cases} \tag{2.15}$$

if, for any $t \geq 0$, it is equal (with probability 1) to the corresponding Itô process,

$$x_t = \mathbf{x}_0 + \int_0^t b(s, x_s)\, ds + \int_0^t \sigma(s, x_s)\, dW_s. \tag{2.16}$$

The following theorem gives conditions for the existence and uniqueness of the strong solution of an SDE.

Theorem 2.3 (Existence and uniqueness) [7] *Suppose that there exists $K > 0$ such that*

$$|\sigma(t, x) - \sigma(t, x')| \le K|x - x'|,$$
$$|b(t, x) - b(t, x')| \le K|x - x'|,$$
$$|\sigma(t, x)| + |b(t, x)| \le k(1 + |x|).$$

Then the stochastic differential equation (2.15) has exactly one strong solution.

When one studies a control problem formulated using SDEs, one needs to estimate some objective functions that depends on the diffusion trajectories and the control dynamics. Therefore, only the distribution of the coupled (diffusion, control) dynamics is relevant. Then a formulation of the solutions of SDEs, which would deploy only the distributions could be more appropriate for control purposes. So, the idea to use 'solutions in distribution' or *weak solutions* might be a good alternative in control problems.

Definition 2.4 (SDE weak solution) We say that a process (x_t) with sample paths in $C_{\mathbb{R}^d}[0, \infty)$ is a weak solution of the stochastic differential equation (2.15) if and only if there exist a probability space $(\Omega, \mathscr{F}, \mathbb{P})$ and stochastic processes $(\widetilde{x}_t)$ and $(\widetilde{W}_t)$ adapted to a filtration $\{\mathscr{F}_t\}$ such that $(\widetilde{x}_t)$ has the same distribution as (x_t), $(\widetilde{W}_t)$ is an $\{\mathscr{F}_t\}$-Wiener process, and

$$\widetilde{x}_t = \widetilde{x}_0 + \int_0^t b(s, \widetilde{x}_s)\, ds + \int_0^t \sigma(s, \widetilde{x}_s)\, d\widetilde{W}_s.$$

Note that the weak solution is the whole element $(\Omega, \mathscr{F}_t, \mathbb{P}, \widetilde{W}_t, \widetilde{x}_t)$. Of course, any strong solution to an SDE is a weak solution: the filtered probability space and the Wiener process remain the same.

Theorem 2.5 (Existence of weak solutions) [217] *Suppose that b and σ are bounded measurable functions, continuous in x for every $t \ge 0$. Then for any $\mathbf{x}_0 \in \mathbb{R}^d$, there exists a weak solution to Eq. (2.15).*

2.3.3 Discontinuous Processes

Markov Jump Processes   A point process over the half-line $[0, \infty)$ is a strictly increasing sequence $(T_n)_{n \ge 1}$ of positive random variables defined on a measurable space $(\Omega, \mathscr{F})$. A *multivariate point process* (also called *marked point process*) is a point process (T_n) for which a random variable X_n is associated to each T_n. The variables X_n take their values in a measurable space $(\mathbf{X}, \mathscr{B})$. With this definition, we may have a finite accumulation point $T_\infty = \lim T_n$, but no point after T_∞.

Intuitively speaking, Markov jump processes are CTMCs defined on continuous state spaces. Any jump process (in the sense of [33, 64]) has embedded a marked point process. These processes were defined first in [33] and called *regular step processes*. Formally, *Markov step processes* are defined as follows.

Consider a measurable space $(\mathbf{X}, \mathscr{B})$ and assume that $\{x\} \in \mathscr{B}$ for all $x \in \mathbf{X}$. The basic ingredients from which we will construct the process are:

1. a nonnegative bounded measurable function λ on $\mathbf{X}$;
2. a stochastic kernel $\mu(x, \Gamma)$ on $\mathbf{X} \times \mathscr{B}(\mathbf{X})$ satisfying $\mu(x, \{x\}) = 0$ for all $x \in \mathbf{X}$.

Then an intuitive description of the process can be given as follows. A particle starting from a point $x_0 \in \mathbf{X}$ remains there for an exponentially distributed sojourn time T_1 with parameter $\lambda(x_0)$ at which time it 'jumps' to a new position x_1 according to the probability distribution $Q(x_0, \cdot)$. Then it remains in x_1 for a holding time T_2, which is exponentially distributed with parameter $\lambda(x_1)$, but which is conditionally independent, given x_1, of T_1. Then it jumps to x_2 according to $Q(x_1, \cdot)$ and so on. The measurable space of events $(\Omega, \mathscr{F})$ associated to this process can be constructed in a canonical way as the set of process trajectories [33].

The *transition probability function* p_t is time-homogeneous and is given by the following expression (according to [33]):

$$p_t(x, dy) = \mu(x, dy) \exp(-\lambda(x)t), \quad t \geq 0.$$

Jump-Diffusion Processes In this section, we give a short presentation of controlled Markov jump-diffusion processes on a probability space $(\Omega, \mathscr{F}, \mathbb{P})$ equipped with a filtration $\{\mathscr{F}_t\}$ that satisfies the usual assumptions, i.e., $\mathbb{P}$-completeness and right continuity with left-hand limits.

Let denote $X \subseteq \mathbb{R}^k$ the state space and $U \subseteq \mathbb{R}^l$ the metric and compact control space. The evolution of the controlled state process in X is governed by an SDE of the following type:

$$dx_t(\theta_t) = b(t, x_t, \theta_t) + \sigma(t, x_t, \theta_t) dW_t + \int_\Gamma q(t, x_{t-}, \rho, \theta_t) \widetilde{N}(dt, d\rho),$$

$$(2.17)$$

where (W_t) denotes a k-dimensional Brownian motion and

$$\widetilde{N}(\omega, t, A) := N(\omega, t, A) - h(t, A)$$

is a compensated homogeneous $\mathscr{F}_t$-Poisson random measure [83] on $\mathbb{R} \times \mathbb{R}^k$ with deterministic compensator

$$h(t, A) = \lambda t \times Q(A)$$

for any $(t, A) \in \mathscr{B} \times \mathscr{B}^k$ and fixed $\lambda \in \mathbb{R}$.

The bounded measurable function

$$q(t, x, \rho, \theta) : \mathbb{R} \times X \times \mathbb{R}^k \times U \to \mathbb{R}^k$$

computes the state- and control-dependent jump size. The Lévy measure $Q(x)$ is assumed to have a compact support $\Gamma \subset \mathbb{R}^k$. If the closure of $A \in \mathscr{B}^k$ does not contain point 0, then $N(t, A) < \infty$ with probability 1.

The meaning of the measure $N(\cdot, t, A)$ is as follows: Observe the random variable

$$\mu(t) = \int_{\mathbb{R}^k} u N(t, du).$$

Then $N(\cdot, t, A)$ is equal to the number of jumps of the process $\mu(t)$ with values in the set A, i.e., the number of instances of time s, $s < t$, such that $\mu(s) - \mu(s^-) \in A$.

The jump term may be alternatively defined as follows.

Consider the process:

$$\tilde{N}_t := \int_0^t \int_\Gamma N(ds, d\rho) = \int_\Gamma N(t, d\rho).$$

This is a Poisson counting process with parameter λ and arrival times $T_i = \inf\{t \mid \tilde{N}_t = i\}$. The interoccurrence times $\{T_{i+1} - T_i\}$ are exponentially distributed with mean value $1/\lambda$. Furthermore,

$$\rho_i = \int_{T_{i-1}}^{T_i} \int_\Gamma \rho N(dt, d\rho)$$

is a sequence of i.i.d. random variables with distribution Q and $\{T_{i+1} - T_i, \rho_i, i < \infty\}$ are mutually independent. Hence, the Poisson measure can be written

$$N(\omega, dt, d\rho) = \sum_{T_i(\omega)} \delta_{(T_i(\omega), \rho_i(\omega))}(dt, d\rho)$$

where δ denotes the Dirac measure. Then

$$\int_{[0,t] \times \Gamma} q(s, x_{s^-}, \rho, \theta_s) N(ds, d\rho) = \sum_{T_i \leq t} q(T_i, x_{T_i^-}, \rho_i, \theta_{T_i}).$$

To ensure the existence of the (stochastic) integrals and the existence and uniqueness of a solution of (2.17), we need the following conditions, which we will call the *standard assumptions* in the sequel.

For any $\theta \in U$ we assume the following conditions (*standard assumptions*): There exist constants $C, L \in \mathbb{R}$ such that

$$\int_0^T |b(s, x(\cdot), \theta)|^2 \, dt < \infty \tag{2.18}$$

$$\int_0^T \int_\Gamma |q(t, x(\cdot), \rho, \theta)|^2 Q(d\rho) \, dt < \infty \tag{2.19}$$

$$|b(t, x, \theta)|^2 + |\sigma(t, x, \theta)|^2 + \lambda \int_\Gamma |q(t, x, \rho, \theta)|^2 Q(dp) \leq C(1 + |x|^2) \tag{2.20}$$

$$|b(t, x_1, \theta) - b(t, x_2, \theta)| \leq L|x_1 - x_2| \tag{2.21}$$

$$\left|\sigma(t,x_1,\theta) - \sigma(t,x_2,\theta)\right|^2 + \lambda \int_\Gamma \left|q(t,x_1,\rho,\theta) - q(t,x_2,\rho,\theta)\right|^2 Q(d\rho)$$

$$\leq L^2 |x_1 - x_2|^2. \tag{2.22}$$

Theorem 2.6 (Existence and uniqueness) *If the functions $b(t,x,\theta)$, $\sigma(t,x,\theta)$ and $q(t,x,\rho,\theta)$ are linearly bounded by the constant C and satisfy a uniform Lipschitz condition with the constant L, i.e., they satisfy the standard assumptions, then the SDE (2.17) has for every $\theta \in U$ a unique solution $x_t \in \Phi$, where $\Phi :=$ $\Phi^k(\mathscr{F}_t, [0,T])$ denotes the class of random processes $\phi(t)$, $t \in [0,T]$, adapted to the filtration $\mathscr{F}_t$ with values in $\mathbb{R}^k$ and sample paths that are right continuous and have left-hand limits.*

If, for any $t_1 \in [0,T)$ and $x \in X$, Eq. (2.17) possesses a unique solution $(x_s^{t_1|x})_{s\in[t_1,T]}$, satisfying the initial condition $x_{t_1}^{t_1|x} = x$, then the family $\{x_s^{t_1|x}, s \in [t_1,T], (t_1,x) \in [0,T) \times X\}$ is a Markov process with the transition kernel

$$p(t,x,s,B) = \mathbb{P}\{x_s^{t_1|x} \in B\}, \quad B \in \mathscr{B}(X).$$

For any transition kernel we have the following requirements:

- $p(t,x,s,\cdot)$ is a probability measure for all fixed $(t,x,s) \in [0,T] \times X \times [0,T]$;
- $p(t,\cdot,s,B)$ is $\mathscr{B}(X)$-measurable for all fixed $(t,s,B) \in [0,T] \times [o,T] \times \mathscr{B}(X)$;
- $p(t,x,t,\{x\}) = 1$;
- Chapman–Kolmogorov equation

$$p(t,x,s,B) = \int_X p(r,y,s,B)p(t,x,r,dy)$$

holds for $t < r < s \in [0,T]$.

2.4 Markov Process Characterisations

In this section, we provide operator concepts, methods and equations that can be used when studying continuous Markov processes.

2.4.1 Operator Methods

Operator methods begin with a local characterisation of the Markov process dynamics. This local specification takes the form of an *infinitesimal generator*. The infinitesimal generator is itself an operator mapping test functions into other functions. From the infinitesimal generator, one can construct a family (semigroup) of conditional expectation operators. The operators exploit the time-invariant Markov structure. Each operator in this family is indexed by the forecast horizon, the interval of time between the information set used for prediction and the object that is being predicted. Operator methods allow us to ascertain global, and in particular, long-run implications from the local or infinitesimal evolution.

The stochastic analysis identifies concepts (like infinitesimal generator, semi-group of operators, resolvent of operators) that characterise in an abstract sense the evolution of a Markov process. Under standard assumptions, all these concepts are equivalent in the sense that, given one concept, all others can be constructed from it. For a detailed presentation of these notions and the connections between them, the reader can consult, for example, [176].

Transition Semigroup of Operators   A one-parameter family of linear operators in a Banach subspace of $\mathscr{B}_b(\mathbf{X})$, $\{\mathbf{P}_t | t \geq 0\}$ is called a *contraction semigroup* if

(1) $\mathbf{P}_0 = I$ (the identity),
(2) $\mathbf{P}_{t+s} = \mathbf{P}_t \mathbf{P}_s$ for all $t, s \geq 0$,
(3) $\|\mathbf{P}_t\| \leq 1$.

Let $\mathscr{B}_0$ be the subset of $\mathscr{B}_b(\mathbf{X})$ consisting of those bounded, measurable functions f for which

$$\lim_{t \searrow 0} \|\mathbf{P}_t f - f\| = 0.$$

The semigroup is called *strongly continuous* on $\mathscr{B}_0$. $\mathscr{B}_0$ is a closed linear subspace of $\mathscr{B}_b(\mathbf{X})$.

Let p_t be a homogeneous transition function. For each $t \geq 0$, define conditional expectation operator by

$$\mathbf{P}_t f(x) := \int f(y) p_t(x, dy) = \mathbb{E}_x f(x_t), \quad x \in \mathbf{X}, \tag{2.23}$$

where $\mathbb{E}_x$ is the expectation with respect to $\mathbb{P}_x$. The Chapman–Kolmogorov equation guarantees that the linear operators $\mathbf{P}_t$ satisfy

$$\mathbf{P}_{t+s} = \mathbf{P}_t \mathbf{P}_s.$$

This suggests another parameterisation for Markov processes: the semigroup of (conditional expectation) operators $\mathscr{P} = (\mathbf{P}_t)_{t>0}$. The *operator semigroup* associated to M maps $\mathscr{B}^b(\mathbf{X})$ into itself. The semigroup $\mathscr{P} = (\mathbf{P}_t)_{t>0}$ can be thought of as an *abstraction* of M, since from $\mathscr{P}$ one can recover the initial process [33]. This kind of abstraction can be related with the concept of abstract control system from [220], but in our case, due to the stochastic features of the model, the domain of the abstraction is no longer the state space $\mathbf{X}$, but $\mathscr{B}^b(\mathbf{X})$.

Example 2.7 (Gaussian semigroup) The transition semigroup of the standard Wiener process on $\mathbb{R}^d$ is given by the formula

$$\mathbf{P}_t(x, dy) = (2\pi t)^{-d/2} \exp\left[-\frac{|y - x|^2}{2t}\right] dy,$$

where dy denotes Lebesgue measure on $\mathbb{R}^d$.

Resolvent of Operators   Recall that the Laplace transform of a function $f : \mathbb{R} \to \mathbb{R}$ is another function, F, defined by

$$F(\alpha) := \int_0^\infty e^{-\alpha t} f(t)\, dt, \quad \alpha > 0. \tag{2.24}$$

Laplace transforms arise in many contexts (linear systems theory, solving ODEs, etc.), one of which is the *moment-generating* functions in basic probability theory. If Y is a real-valued random variable with probability law $\mathbb{P}$, then the moment-generating function is

$$M_Y(\alpha) := \mathbb{E}\big(e^{\alpha Y}\big) = \int e^{\alpha Y}\, d\mathbb{P} = \int e^{\alpha y} p(y)\, dy$$

when the density in the last expression exists. You may recall that the distributions of well-behaved random variables are completely specified by their moment-generating functions. In fact, this is a special case of a more general result about when functions are uniquely described by their Laplace transforms, i.e., when f can be expressed uniquely in terms of F.

This is important for the transition semigroup, since it turns out that the Laplace transform of a semigroup of operators is better behaved than the semigroup itself, and it is more closely connected with the properties of the trajectories. It might be important to say when we can use the Laplace transform to recover the semigroup.

Because the Laplace transform of a function is again a function according to (2.24), the Laplace transform of the transition operator $\mathbf{P}_t$ will be again an operator. The *resolvent of operators* $\mathscr{V} = (V_\alpha)_{\alpha \geq 0}$ associated with the semigroup $\mathscr{P}$ is given by the formula

$$V_\alpha f(x) := \int_0^\infty e^{-\alpha t} \mathbf{P}_t f(x)\, dt. \tag{2.25}$$

Let V be the initial operator V_0 of $\mathscr{V}$; V is known as the *kernel operator* of Markov process M.

When the function f is a value (like cost, utility) function, $\mathbf{P}_t f(x)$ is the expected value at time t when the process starts in the state x. $V_\alpha f(x)$ can be thought of as the net present expected value when the process starts at x and we apply a discount rate α.

Infinitesimal Generators   Associated with the semigroup $(\mathbf{P}_t)$ is its *strong generator or infinitesimal generator* which, loosely speaking, is the derivative of $\mathbf{P}_t$ at $t = 0$. Let $D(L) \subset \mathscr{B}^b(\mathbf{X})$ be the set of functions f for which the following limit exists:

$$\lim_{t \searrow 0} \frac{1}{t}(\mathbf{P}_t f - f) \tag{2.26}$$

and denote this limit Lf. The limit refers to convergence in the norm $\|\cdot\|$, i.e., for $f \in D(L)$ we have

$$\lim_{t \searrow 0} \left\| \frac{1}{t}(\mathbf{P}_t f - f) - Lf \right\| = 0.$$

Specifying the domain $D(L)$ is essential when we define the operator L.

Some simple properties of the infinitesimal generator are:

- the generator L of each semigroup of homogeneous transition operators $(\mathbf{P}_t)$ is a linear operator;
- if L is the generator of the semigroup $(\mathbf{P}_t)$, and f is in the domain of L, then $\mathbf{P}_t$ and L commute, for all t, that is, $\mathbf{P}_t L f = L \mathbf{P}_t f$.

If $(\mathbf{P}_t)$ is a strongly continuous contraction semigroup, then $D(L)$ is dense. In addition, L is closed, that is, if $f_n \in D(L)$ converges to f and $L f_n$ converges to g then $f \in D(L)$ and $L f = g$. If $(\mathbf{P}_t)$ is a strongly continuous contraction semigroup we can reconstruct $\mathbf{P}_t$ using its infinitesimal generator L (e.g., [83] Proposition 2.7 of Chap. 2). This suggests using L to parameterise the Markov process. The Hille–Yosida theorem (e.g., [83], Theorem 2.6 of Chap. 1) gives necessary and sufficient conditions for a linear operator to be the generator of a strongly continuous positive contraction semigroup. Necessary and sufficient conditions to ensure that the semigroup can be interpreted as a semigroup of conditional expectations are also known (e.g., [83], Theorem 2.2 of Chap. 4).

Example 2.8 (Generator of a Markov jump process) Let (x_t) be a Markov jump process on $(\mathbf{X}, \mathscr{B})$, with $\mu(x, \Gamma)$ as transition function and λ its transition rate. Then

$$Lf(x) = \lambda(x) \int \big[f(y) - f(x) \big] \mu(x, dy) \qquad (2.27)$$

defines on $\mathscr{B}_b(\mathbf{X})$ a bounded linear operator L, which is the generator of (x_t).

Example 2.9 (Generator of a diffusion process) Let (x_t) be a diffusion process on $\mathbb{R}^d$ generated by an SDE like

$$\begin{cases} dx_t = b(x_t) + \sigma(x_t)\, dW_t \\ x_0 = \mathbf{x}_0 \end{cases} \qquad (2.28)$$

with the drift coefficient b and the diffusion coefficient σ. Then

$$Lf(x) = \sum_i b_i(x) \frac{\partial}{\partial x_i} f(x) + \frac{1}{2} \sum_{i,j} a_{ij}(x) \frac{\partial^2}{\partial x_i \partial x_j} f(x), \qquad (2.29)$$

where $A := \sigma \sigma^{\mathsf{T}} = (a_{ij})$. Here, σ^{T} denotes the transpose matrix of a matrix σ. The domain $D(L)$ is given by $C_c^2(\mathbb{R}^d)$, the twice continuously differentiable functions with compact support in $\mathbb{R}^d$.

Excessive Functions   A function f is *excessive* (with respect to the semigroup $(\mathbf{P}_t)$ or the resolvent (V^α)) if it is measurable, nonnegative and

$$\begin{aligned} \mathbf{P}_t f \leq f \quad &\text{for all } t \geq 0, \\ \mathbf{P}_t f \nearrow f \quad &\text{as } t \searrow 0. \end{aligned} \qquad (2.30)$$

Let $\mathscr{E}_M$ be the set of all excessive functions associated to M. For any $\alpha > 0$, a measurable, nonnegative function f is said to be *α-excessive* if $e^{-\alpha t} \mathbf{P}_t f \nearrow f$ as

$t \searrow 0$. A typical example of an α-excessive function is $u = V_\alpha f$ for nonnegative measurable f.

The excessive functions play the role of the super-harmonic functions from the theory of partial differential equations. Moreover, over the process trajectories, these functions behave like the Lyapunov functions in the deterministic case. This is the reason why these functions are also called *stochastic Lyapunov functions* (mostly in the stochastic control literature).

The strong Markov property can be characterised in terms of excessive functions [183, 208].

Quadratic Forms   Suppose $D = L^2(\mathbf{X}, \mu)$ (the space of square integrable μ-measurable extended real-valued functions on $\mathbf{X}$), where we have the natural inner product

$$\langle f, g \rangle = \int f(x)g(x) \, d\mu(x). \tag{2.31}$$

If $f \in \mathscr{D}(L)$ and $g \in L^2(\mu)$ then we may define the (closed) form

$$\mathscr{E}(f, g) = -\langle Lf, g \rangle. \tag{2.32}$$

This leads to another way of parameterising Markov processes. Instead of writing down a generator one starts with a form. As in the case of a generator it is typically not easy to fully characterise the domain of the form. For this reason one starts by defining a form on a smaller space and showing that it can be extended to a closed form in subset of $L^2(\mu)$. When the Markov process can be initialised to be stationary, the measure μ is typically this stationary distribution. More generally, μ does not have to be a finite measure.

This approach to Markov processes was pioneered by Beurling and Deny [23] and Fukushima [93] for symmetric Markov processes. In this case both the operator L and the form $\mathscr{E}$ are symmetric. A stationary symmetric Markov process is time reversible. If time were reversed, the transition operators would remain the same. On the other hand, multivariate standard Brownian motion is a symmetric nonstationary Markov process that is not time reversible. The literature on modelling Markov processes with forms has been extended to the nonsymmetric case by Ma and Rockner [176].

2.4.2 Kolmogorov Equations

We have seen that, for CTMCs, the connection between transition probabilities and infinitesimal generators is described by some ODEs called *Kolmogorov forward–backward equations*. Naturally, such equations are extended for continuous time continuous space Markov processes. Since the infinitesimal generator for general Markov processes may be a linear operator that can take different expressions (differential/integral operator/integro-differential operator), the equations are more than some simple ODEs (such as in the case of CTMCs). However, they provide powerful insights on the inherent links between the transition probabilities and the infinitesimal generator.

Proposition 2.10 [64]

(i) *If* $f \in \mathscr{B}_0$ *and* $t \in \mathbb{R}_+$, *then* $\int_0^t \mathbf{P}_s f \, ds \in D(L)$ *and*

$$L\left(\int_0^t \mathbf{P}_s f \, ds\right) = \mathbf{P}_t f - f.$$

(ii) *If* $f \in D(L)$ *and* $t \in \mathbb{R}_+$, *then* $\mathbf{P}_t f \in D(L)$ *and*

$$\frac{d}{dt}\mathbf{P}_t f = L\mathbf{P}_t f = \mathbf{P}_t Lf. \tag{2.33}$$

(iii) *If* $f \in D(L)$ *and* $t \in \mathbb{R}_+$, *then* $\mathbf{P}_t f \in D(L)$ *and*

$$\mathbf{P}_t f - f = \int_0^t L\mathbf{P}_s f \, ds = \int_0^t \mathbf{P}_s Lf \, ds. \tag{2.34}$$

The equalities (2.33) represent the Kolmogorov backward–forward equations associated to a Markov process. The equalities (2.34) are the integral representations of such equations. Moreover, for practical calculations, (2.34) can be rewritten as

$$\mathbb{E}_x f(x_t) = f(x) + \mathbb{E}_x \int_0^t L\mathbf{P}_s f \, ds, \quad f \in D(L). \tag{2.35}$$

This is known as the *Dynkin formula* and it represents the basis for studying the connection between Markov processes and martingales, which is given next.

2.4.3 Martingale Problem and Extended Generator

One approach towards obtaining strong Markov processes is through martingales, and more specifically through the martingale problem.

Definition 2.11 (Martingale problem) Let $\mathbf{X}$ be a Polish space, $\mathscr{D}$ a class of bounded, continuous, real-valued functions on $\mathbf{X}$, and L an operator from $\mathscr{D}$ to bounded, measurable functions on $\mathbf{X}$. An $\mathbf{X}$-valued stochastic process (x_t) on $[0, +\infty)$ is a solution to the martingale problem for L and $\mathscr{D}$ if, for all $f \in \mathscr{D}$,

$$f(x_t) - \int_0^t Lf(x_s) \, ds \tag{2.36}$$

is a martingale with respect to the natural filtration of (x_t).

Suppose (x_t) is a càdlàg solution to the martingale problem for L, $\mathscr{D}$. Then for any $f \in \mathscr{D}$, the stochastic process given by (2.36) is also càdlàg.

Let $M = (\Omega, \mathscr{F}, \mathscr{F}_t, x_t, (\mathbb{P}_x)_{x \in X})$ be a Markov process, with the infinitesimal generator L. For $f \in D(L)$ we define the real-valued process $(C_t^f)_{t \geq 0}$ by

$$C_t^f = f(x_t) - f(x_0) - \int_0^t Lf(x_s) \, ds. \tag{2.37}$$

Proposition 2.12 (Markov processes solve martingale problems) [64] *If M has càdlàg sample paths and $f \in \mathscr{D}$, where $\mathscr{D}$ represents the set of continuous functions from $D(L)$, then M solves the martingale problem for L and $\mathscr{D}$.*

Proposition 2.13 (Solutions to the martingale problem are strongly Markovian) *Suppose that for each $x \in X$, there is a unique càdlàg solution to the martingale problem for L, $\mathscr{D}$ such that $x_0 = x$ (i.e., the process $(C_t^f)_{t \geq 0}$ is a martingale on $(\Omega, \mathscr{F}, \mathscr{F}_t, \mathbb{P}_x)$). Then the collection of these solutions is a homogeneous strong Markov family M, whose generator is equal to L on $\mathscr{D}$.*

There may be other functions f, not in $D(L)$, for which something akin to (4.21) is still true. In this way, we obtain an operator that extends the infinitesimal generator to a larger domain. This is called the *extended generator* of the process, and its formal definition is given below.

Let $D(\widehat{L})$ denote the set of measurable functions $f : \mathbf{X} \to \mathbb{R}$ with the following property: There exists a measurable function $h : \mathbf{X} \to \mathbb{R}$ such that the function $t \to h(x_t)$ is integrable $\mathbb{P}_x$-a.s. for each $x \in \mathbf{X}$ and the process

$$C_t^f = f(x_t) - f(x_0) - \int_0^t h(x_s) \, ds \tag{2.38}$$

is a local martingale. Then we write $h = \widehat{L}f$ and $(\widehat{L}, D(\widehat{L}))$ represents the extended generator of the process (x_t).

We can define the extended generator using the operator resolvent $(V_\alpha)_{\alpha > 0}$ associated to the transition semigroup $(\mathbf{P}_t)_{t \geq 0}$. A function $f \in \mathscr{B}^b(\mathbf{X})$ is said to belong to the domain of the extended generator of $(\mathbf{P}_t)_{t \geq 0}$ if there exists a Borel measurable function h such that $V_\alpha |h| < \infty$ and

$$f = V_\alpha(\alpha f - h) \tag{2.39}$$

for some $\alpha > 0$ or for all $\alpha > 0$ (see [35]). It is easy to prove that the property (2.39) is equivalent to the martingale property of (2.38). The function h is unique up to changes on a set of potential zero. $A \in \mathscr{B}$ is of potential zero if

$$V_\alpha 1_A = 0$$

for some $\alpha > 0$ or for all $\alpha > 0$. Explicitly, h can be given as

$$h = \lim_{n \to \infty} n(f - nV_n f).$$

The importance of the martingale problem is illustrated by the connection between the solutions of the martingale problem and the weak solutions of the stochastic integro-differential equations that correspond to Markov processes. To understand this, we need to discuss these connections first for diffusion processes. The key point is the fact that a solution for the martingale problem corresponding to a diffusion process gives a weak solution of the associated Itô equation. This is a classical result, which is well known in the stochastic literature, since the seminal work of Stroock and Varadhan [217]. The result can be formulated as follows.

Theorem 2.14 *(x_t) is a solution with paths in $C_{\mathbb{R}^d}[0, \infty)$ of the martingale problem for L given by (2.29) and $\mathscr{D} = C_c^2(\mathbb{R}^d)$ if and only if (x_t) is a weak solution for (2.28).*

In fact, for the diffusion defined by (2.28), the martingale associated can be derived using the Itô formula and the properties of the Itô integral

$$f(x_t) - f(x_0) - \int_0^t Lf(x_s)\,ds = \int_0^t \nabla f(x_s)^{\mathsf{T}} \sigma(x_s)\,dW_s. \qquad (2.40)$$

Taking expectations in (2.40), we obtain the identity

$$v_t f = v_0 f + \int_0^t v_s Lf\,ds, \qquad (2.41)$$

which represents the weak form of the forward equation for the one-dimensional diffusion distributions denoted by $\{v_t\}$. Moreover, the converse of the fact that every solution of the martingale problem gives a solution of the forward equation is true as well, and the following theorem can be proved.

Theorem 2.15 *If (x_t) is a solution of the martingale problem for L defined by (2.29), then $\{v_t\}$, the one-dimensional distributions of (x_t), is a solution of the forward equation (2.41). If $\{v_t\}$ is a solution of the forward equation (2.41), then there exists a solution of the martingale problem for L whose one-dimensional distributions are given by $\{v_t\}$.*

Theorem 2.15 shows that the standard approaches for specifying diffusion processes (stochastic differential equations, martingale problems, and forward equations) are, under general conditions, equivalent in the sense that the existence of one implies the existence of the other two. The same argument applies for the uniqueness of the solutions for such problems. For general Markov processes that solve some integro-differential equations, these approaches can be leveraged and new characterisations can be proved. The price to pay is that it requires the construction of a Poisson random measure from the sample paths of the solution of the martingale problem. The interested reader may consult the newly appeared article on this subject of Thomas G. Kurtz [155].

2.5 Some Remarks

To understand this book, one needs a course in applied probability and Markov processes, from discrete to continuous time and space. In standard textbooks, treatments of Markov process theory tend to concentrate more on the countable state Markov chains (from the 'applied probability' perspective), or on purely continuous space Markov processes (from the 'pure mathematics' perspective).

The main goal of this book is to study reachability problems for stochastic hybrid processes. These processes encompass a hybrid continuous/discrete dynamics. By 'zooming in' closely, then 'zooming out' for an overview of the investigation, we are able to describe the dynamics through discrete and continuous space Markov processes. In this chapter, we give a brief overview of such processes. Moreover, for each type of Markov model, we present the characterisation and tooling apparatus that can be employed in theoretical and practical applications. The guiding thread of this presentation is provided by the infinitesimal generator of the Markov process, which is a matrix for the discrete case and a linear operator for the continuous case. All other concepts (transition probabilities/semigroup, resolvent, extended generator, martingale characterisation) are strongly related to this infinitesimal generator. Furthermore, we will see in this book that most of the computational methods associated to different features (control, optimisation, verification, performance, etc.) are based on a handling of the properties of this infinitesimal generator.

Chapter 3
Hybrid Systems: Deterministic to Stochastic Perspectives

3.1 Overview

The multi- and interdisciplinary research field of hybrid systems has emerged over the last two decades and lies at the boundary between computer science, control engineering and applied mathematics. In general, a hybrid system can be defined as a system built from atomic discrete components and continuous components by parallel and/or serial composition, arbitrarily nested. The behaviours and interactions of components are governed by models of computation. Hybrid phenomena captured by such mathematical models are manifested in a great diversity of complex engineering applications such as real-time systems, embedded software, robotics, mechatronics, aeronautics and process control.

The high-profile and safety-critical nature of such applications has fostered a large and growing body of work on

- formal methods for hybrid systems: mathematical logic, computational models and methods and automated reasoning tools supporting the formal specification and verification of performance requirements for hybrid systems, and
- the design and synthesis of control programs for hybrid systems that are provably correct with respect to formal specifications.

Reachability is a fundamental concept in the study of dynamical systems in general and hybrid systems in particular. The reachable set consists of all the states that can be visited by a trajectory of the hybrid system starting in a prespecified set of initial states (see Fig. 3.1). In many cases, reachability analysis has been motivated by the safety verification problem, which consists of checking whether the intersection of the reachable set with a set of bad/dangerous states is empty. In addition, reachability analysis can be used for studying performance/performability properties of the given system.

This chapter provides an incremental perspective, from deterministic to stochastic, of the modelling and analysis approaches existing for hybrid discrete/continuous systems. We start with deterministic hybrid systems and their associated problems and end up with different models for uncertainty models of hybrid systems that

L.M. Bujorianu, *Stochastic Reachability Analysis of Hybrid Systems*,
Communications and Control Engineering,
DOI 10.1007/978-1-4471-2795-6_3, © Springer-Verlag London Limited 2012

Fig. 3.1 Reachability problem

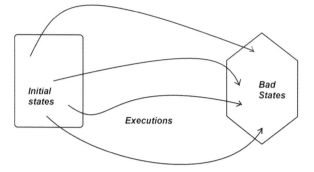

result by randomisation of the deterministic ones. Specific modelling and analysis problems of the deterministic hybrid systems are briefly described and discussions regarding the leveraging of these problems to the stochastic case will follow.

Objectives   The objectives of this chapter can be summarised as follows:

1. To provide a smooth introduction in the area of hybrid system modelling; important classes of models for hybrid systems are underlined.
2. To explain how and where the randomness arises using deterministic modelling paradigms for hybrid systems: We describe how these 'classical' models can be enhanced with uncertainty measures and transformed into new models that require sometimes completely different tools for analysis.
3. To present analysis methods for deterministic hybrid systems and discuss how these have been extended to the probabilistic framework.

3.2 Deterministic Hybrid Systems

In a general sense, hybrid systems are dynamical systems that contain both discrete and continuous variables. Discrete variables are variables that can only take values in a certain countable set, like

$$\mathbb{N} = \{0, 1, 2, 3, \ldots\}$$

or

$$Q = \{q_{00}, q_{01}, q_{10}, q_{11}\}.$$

On the other hand, continuous variables can take values in some set of a Euclidean space. Discrete and continuous dynamics do not only coexist in the same system, but they may also interact, in the sense that continuous variables may influence the discrete variable dynamics and vice versa. Hybrid systems have been the focus of intense research since the early 1990s by control theorists, computer scientists and applied mathematicians. The main motivation for the research in this area has been the importance of hybrid systems in a range of applications, especially in the area of embedded computation and control systems.

Hybrid phenomena arise, in many cases, from the internal actions of the system (plant) or from the action of an external device like a controller or a supervisor. The latter includes, usually, human intervention that involves discrete actions in the form of decision making. Examples of internally generated hybrid phenomena include disturbances, failures and operating mode changes. There are many examples of externally generated hybrid behaviours such as control of a device with a finite number of states (e.g., 'on'/'off') and controlled switching between control laws. External hybrid phenomena may appear in networks due to the interaction between subsystems.

A very common example is that of a thermostat in a room. The temperature in the room is a continuous variable and the state of the thermostat ('on' or 'off') is discrete; the continuous and discrete parts cannot be described independently, since, clearly, there is interaction between the two.

Hybrid systems theory has roots in several different fields of science. Within control theory, we have to mention several lines of research that indeed constituted the motivation for the study of hybrid systems. Work in non-linear control theory has driven researchers to the insight that switching control is somehow fundamentally more powerful than smooth control. Adaptive control developments have naturally led to the consideration of switching control schemes as well. Moreover, it is known that various types of switching have traditionally been used in control methods and implementations such as gain scheduling, sliding mode control and programmable logic controllers. Also, fuzzy control is in this line of research, since fuzzy schemes are typically based on a combination of different operating regimes.

Outside of control theory, computer scientists have been notably active in promoting hybrid systems theory. Researchers in software verification have been working towards finding a more precise description of the processes in which computers play a role. Therefore, the motivation has been to admit continuous variables and differential equations into their studies. The interests of computer scientists have brought new research problems, for instance, ones concerning safety verification and validation, which have received less attention from control theorists. These questions engendered renewed interest in some classical control topics such as reachability analysis.

The mathematical treatment of hybrid systems is interesting in that it is based on the framework of non-linear control, but its mathematics is qualitatively distinct from the mathematics of purely discrete or purely continuous phenomena. Methods and problems from dynamical system theory have raised new research problems when they have been extended to hybrid dynamical systems.

3.2.1 Modelling

Hybrid models can be divided into the following classes: automata and transition systems, ordinary differential equations, algebraic approach and programming languages. By far, the first two classes are the most common approaches used for modelling hybrid systems. These two classes differ in the classification methodology of the dynamics: discrete (e.g., automata) versus continuous (namely, ordinary differential equations). The focus of the third class is to establish an appropriate algebraic formalism for hybrid behaviours. The fourth class is completely different from the others, since the hybrid behaviour is described by a programming language via some primitive instructions and rules to combine these instructions. These last two classes are not of interest for this monograph.

Adequate models for hybrid systems are obtained by considering finite control location graphs supplied with discrete data structures (counters, stacks, etc.) and real-valued variables that change continuously at each control location. The transitions between control locations are conditioned by constraints on the values (or configurations) of the (discrete and/or continuous) variables and data structures of the system; the execution of these transitions resets the discrete data structures and the continuous variables of the system.

Automata Perspective   The most often used formal models for hybrid systems are the hybrid automata, which have proved to be very useful for developing methods for hybrid control law design, simulation and verification. In particular, hybrid automata integrate diverse models such as differential equations and state machines in a single formalism with a uniform mathematical semantics and novel algorithms for safety and real-time performance analysis.

First introduced in [115], hybrid automata represent a formal model that combines discrete control graphs, usually called finite state machine or automata, with continuously evolving variables. A hybrid automaton exhibits two kinds of state change:

1. discrete jump transitions, which occur instantaneously, and
2. continuous flow transitions, which occur when time elapses.

The structure of the hybrid automaton model is made up of two parts:

1. a finite state machine (FSM), which controls the model's 'modes of operation', and
2. a collection of alternate dynamical models, which includes a collection of dynamical state model variables.

The FSM is a transition model that consists of a collection of 'states' and a definition of transitions among those states, defined by:

1. the 'from state' for the transition;
2. the input that, if received while in the 'from' state, triggers this transition to be performed;
3. the 'to state' for this transition, and
4. the output from the FSM if this transition takes place.

At a particular moment in time, the FSM can only be 'in' one state and the 'from state' and input combination must be unique for the FSM's transition model.

The FSM is commonly represented as a directed connected graph where each node represents a 'state'. Since a single node may have multiple transitions to any

other given node, triggered by distinct inputs, the graph may have many links between any two nodes.

Now we will introduce the formal definitions of a hybrid automaton and its semantics as given in [115] and in many other references related to hybrid systems.
For a modern presentation of such concepts, we use the lecture notes on hybrid
systems of Lygeros [169].

Definition 3.1 (Hybrid automaton) A *hybrid automaton* is a collection

$$H = (Q, \mathscr{X}, \text{Init}, f, \text{Dom}, \mathscr{E}, \mathscr{G}, \mathscr{R})$$

where:

(a) Q is a finite set of discrete locations (modes);
(b) $\mathscr{X} = \mathbb{R}^n$ is a set of continuous states;
(c) Init $\subset Q \times \mathscr{X}$ is a set of initial states;
(d) $f : Q \times \mathscr{X} \to T\mathscr{X}$ is a vector field;
(e) Dom $: Q \to 2^{\mathscr{X}}$ is a (continuous state) domain;
(f) $\mathscr{E} \subset Q \times Q$ is a set of edges;
(g) $\mathscr{G} : \mathscr{E} \to 2^{\mathscr{X}}$ is a guard condition;
(h) $\mathscr{R} : \mathscr{E} \times \mathscr{X} \to 2^{\mathscr{X}}$ is a reset map.

Recall that $2^{\mathscr{X}}$ denotes the power set (all the subsets) of $\mathscr{X}$. In the previous
definition, the function Dom assigns a set $\text{Dom}(q)$ of continuous states to each
discrete mode q. We refer to

$$(q, x) \in Q \times \mathscr{X}$$

as the hybrid state of H. The main assumption is that f is Lipschitz continuous in its
second argument. With each discrete mode $q \in Q$, we associate a set of continuous
initial states $\text{Init}(q) = \{x \in \mathscr{X} : (q; x) \in \text{Init}\}$, a vector field $f(q;)$ and the invariant
set $\text{Dom}(q)$.

The semantics of hybrid automata defines possible evolutions for their state. Informally, starting from an initial state $(q_0, x_0) \in \text{Init}$, we choose a solution of the
ODE:

$$\frac{dx}{dt} = f(q_0, x),$$
$$x(0) = x_0,$$

and follow it (continuous dynamics). While the continuous state moves along this
continuous path, the discrete state remains constant, $q(t) = q_0$. The continuous
dynamics keeps going on as long as x evolves inside the invariant $\text{Dom}(q_0)$. In
case the continuous path reaches the guard $\mathscr{G}(q_0, q_1)$ corresponding to some edge
$(q_0, q_1) \in \mathscr{E}$, the discrete state may change value to q_1. If this discrete transition
takes place, then the continuous state has to be reset with a new value in the set
$\mathscr{R}(q_0, q_1, x) \subset \mathscr{X}$. After that, the continuous evolution restarts and the whole process is repeated.

Fig. 3.2 One-dimensional hybrid dynamics

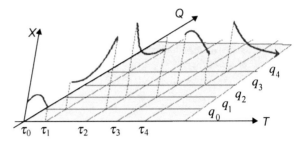

Hybrid automata involve both continuous flow determined by differential equations and discrete transitions determined by an automaton. One way to characterise the dynamics of the state of a hybrid automaton is to consider the set of times that contains both continuous-time intervals (provided by the intervals when the continuous state evolves) and distinguished discrete points in time (provided by moments when the discrete transitions occur). Such a set of times is called a *hybrid time set*.

Definition 3.2 (Hybrid time set) A hybrid time set is a sequence of intervals $\tau = \{I_0, I_1, \ldots, I_N\} = \{I_i\}_{i=0,\ldots,N}$ finite or infinite (i.e., $N = \infty$ is allowed) such that:

(i) $I_i = [\tau_i, \tau_i'] \; \forall i < N$;
(ii) if $N < \infty$ then $I_N = [\tau_N, \tau_N']$ or $I_N = [\tau_N, \tau_N')$;
(iii) $\tau_i \le \tau_i' = \tau_{i+1} \; \forall i$.

Definition 3.3 (Hybrid trajectory) A hybrid trajectory is a triple (τ, q, x) consisting of a hybrid time set $\tau = \{I_i\}_{i=0,\ldots,N}$ and two sequences of functions $q = \{q_i(\cdot)\}_{i=0}^{N}$ and $x = \{x_i(\cdot)\}_{i=0}^{N}$ with $q_i(\cdot): I_i \to Q$ and $x_i(\cdot): I_i \to \mathcal{X}$ (see Fig. 3.2).

An execution of an autonomous hybrid automaton is a hybrid trajectory, (τ, q, x) of its state variables. We have restrictions on the types of hybrid trajectory that the hybrid automaton can accept, as is stated in the following definition of the hybrid execution.

Definition 3.4 (Hybrid automaton execution) An execution of a hybrid automaton H is a hybrid trajectory, (τ, q, x), subject to the following conditions:

(i) initial condition:

$$\big(q_0(0), x_0(0)\big) \in \text{Init};$$

(ii) discrete evolution:

$$\big(q_i(\tau_i'), q_{i+1}(\tau_{i+1})\big) \in \mathcal{E},$$
$$x_i(\tau_i') \in \mathcal{G}\big(q_i(\tau_i'), q_{i+1}(\tau_{i+1})\big),$$
$$q_{i+1}(\tau_{i+1}) \in \mathcal{R}\big(q_i(\tau_i'), q_{i+1}(\tau_{i+1}), x_i(\tau_i')\big);$$

(iii) continuous evolution: while the discrete state remains constant in the interval I_i

$$q_i(t) = q_i(\tau_i) \quad \forall t \in I_i,$$

the continuous state $x_i(\cdot) : I_i \to \mathscr{X}$ is solution to the differential equation

$$\frac{dx_i}{dt} = f\big(q_i(t), x_i(t)\big)$$

over I_i starting at $x_i(\tau_i)$; subject to the constraint

$$x_i(t) \in \mathrm{Dom}\big(q_i(\tau_i)\big) \quad \forall t \in [\tau_i, \tau_i').$$

The statements (i) and (ii) regarding the discrete and continuous evolution are true for all $i = 0, \ldots, N$.

Dynamical System Perspective From the beginning of the theory of hybrid systems, many researchers have attempted to extend the theory of dynamical systems so that it might capture discrete phenomena. The most prominent example in this area is the work of Branicky [38, 39]. Hybrid dynamical systems belong to the systems approach and they involve an interacting countable collection of dynamical systems. The control of such systems has a hierarchical structure given by continuous-time evolutions at lower-level operational modes and logical decision-making units at the higher level of the structure.

Formally, a hybrid dynamical system is an indexed collection of dynamical systems together with a map for 'jumping' among them. The jumping mechanism has to specify the switching between dynamical systems and/or resetting of the state. This jumping occurs whenever the dynamical system state satisfies certain conditions, described by a prespecified subset of the state space. Therefore, the whole system can be described as a sequential piecing out of dynamical systems with initial and final states, in which the switchings reset the initial state for a different dynamical system whenever a final state for the previous dynamical system has been reached.

Hybrid dynamical system behaviours manifest a very rich palette of properties. In particular, they can exhibit multiple complex phenomena such as: Zeno behaviour, discontinuities, deadlock, beating or livelock and confluence or merging of solutions.

A Zeno behaviour is characterised by the apparition of a trajectory with infinitely many resettings (jumps) in finite time. Deadlock corresponds to a dynamical system state (an absorbent/cemetery state) from which no (continuous or discrete) continuation is possible. Beating appears when the system trajectory encounters the same resetting surface a finite or infinite number of times in zero time. Finally, a hybrid dynamical system experiences confluence when some of its trajectories coincide after a certain time instant. These types of phenomenon, together with the loss of important properties of classical dynamical system theory (e.g., continuity of solutions, continuous dependence of solutions on the systems initial conditions, uniqueness of solutions), make the analysis of hybrid dynamical systems an open research area.

3.2.2 Analysis

One research direction to modelling and analysis of hybrid systems is represented by the extension of the techniques that have been developed before for finite state automata to include systems with simple continuous dynamics. This approach, developed mostly in computer science, uses one of two following analysis techniques:

- algorithmic verification (*model checking*), which verifies a system specification symbolically on all system executions; and
- deductive verification (*theorem proving*), which proves a specification by induction on all system executions.

In this context, the key problem entails computability and decidability issues. In other words, the objective is to prove that the question 'Does the system satisfy the specification?' can be answered using an algorithm with a finite number of steps.

Deductive verification analyses system properties through formal deduction based on a set of inference rules. Deductive verification is applicable to infinite state systems but has a drawback in the sense that guidance from a user is almost always needed in the process. Theorem proving for control engineering was introduced and developed by Martin and collaborators (see [36] for an overview). Essentially, specific control system properties are formalised in a kind of model logic and proof support is based on huge continuous mathematics formal theory libraries. This approach, although perfectly legitimated, encountered resistance from control engineers. The method requires very specialised background, complex interdisciplinary collaboration and time-consuming activities.

Model checking is by far more effective and has achieved a relatively wider acceptance in control engineering. In model checking, a property is given as a formula of a temporal logic and automatically compared with a state transition graph representing the actual behaviour of the system. The main advantage of this method is its efficiency: Model checking is linear in the product of the size of the structure and the size of the formula, when the logic is the branching-time temporal logic CTL (computation tree logic) [56]. Most model checkers can be entirely automated and many of them can support system descriptions based on continuous mathematics. Usually, model-checking techniques are applicable to finite state systems and basically perform an exhaustive exploration of all possible system executions in a fully automated way. The major drawback of model checking is the state explosion problem, i.e., the number of system trajectories that need to be explored grows very quickly as the number of states increases. There exist methods, for example the use of an efficient data structure called *ordered binary decision diagrams* [40], that allow model-checking systems with a huge number of states. But when the number of possible states is infinite, as in the case of the continuous state spaces, model checking is no longer applicable. This approach works only in specialised situations, and in many cases infinite state systems do not admit finite abstractions. For hybrid systems, the main difficulty in applying model checking is the continuous part of the system's state space.

Symbolic model-checking procedure and its implementation HYTECH for linear hybrid automata have been developed through manipulating and simplifying $(\mathbb{R}, \leq, +)$-formulae [6, 114]. The underlying system model is a hybrid automaton, an extension of finite automata with continuous variables that are governed by differential equations. The specification language is given by the integrator computation tree logic ICTL, a branching-time logic with clocks and stop watches for specifying timing constraints. This language can be useful for specifying safety, liveliness, real-time and duration requirements for hybrid systems. Provided we have supplied a hybrid automaton describing a system and an ICTL describing a requirement, HYTECH computes the state predicate that characterises the set of system states where the requirement is satisfied.

UPPAAL [111] is another model checker suitable for automatic verification of safety and boundedness liveliness properties of real-time systems modelled as networks of timed automata. Its main characteristics are: (i) a graphical interface that supports graphical and textual representations of networks of timed automata and automatic transformation of graphical representations to textual format; (ii) a compiler that transforms a certain class of linear hybrid systems to networks of timed automata and (iii) a model checker whose implementation is based on constraint solving techniques. UPPAAL also supports diagnostic model-checking providing diagnostic information in case verification of a particular real-time systems fails. UPPAAL allows linear hybrid automata where the speed of clocks is given by an interval. Hybrid automata of this form may be transformed into ordinary timed automata using a specific translator. The latest version of UPPAAL is able to check for reachability properties, in particular, whether certain combinations of control nodes and constraints on clocks and integer variables are reachable from an initial configuration.

Another research direction to modelling and analysis for hybrid systems is represented by the extension of techniques that were developed previously for continuous state space and continuous-time dynamical systems and control [224]. The outline of the research here has been towards extending the standard modelling, reachability and stability analyses and controller design techniques for capturing the interaction between continuous and discrete dynamics [168, 170]. This approach consists of leveraging the existing control techniques (stability theory [38], optimal control [38, 188] and control of discrete event systems [163]) to hybrid systems.

Reachability results exist for simpler hybrid systems (linear hybrid automata, piecewise affine hybrid systems) [156, 157, 223]. In the case of hybrid systems whose dynamics are non-linear or of order greater than one, difficult problems regarding the computable solutions for the reachability problem have risen. Good progress has been made in cases of more general hybrid system and various results have been reported in the literature (see e.g., [62, 184]). In this direction, the most efficient methods are based on optimal control techniques [170, 171]. The problem of computing reachable sets is addressed via optimal control and game theory methods that have been used before for automata and dynamical systems. Based on these, Hamilton–Jacobi equations can be derived and their solutions describe the boundaries of the reachable sets.

For systems with complex continuous dynamics, there exist methods that compute approximations of the reachable sets. For optimal control, these techniques use geometric representations for the reachable sets (including polytopes, ellipsoids, or level set) and approximations of dynamics (linear dynamics). Hence, the optimal control problems are easier to solve. Deductive techniques are inherently based on over-approximation of reachable sets. Model-checking methods use approximations of the system by another simpler system that allows easier computations. The above methods are concerned with explicitly deriving the set of reachable states.

Simulation can also be used for reachability investigations. Simulation is a procedure of generating partial traces by executing the model and then checking the set of partial traces against its specification. This standard approach to verifying certain properties of a hybrid system consists of finding an equivalent transition system, called a *bisimulation*, with a finite number of states. Bisimulations are reachability-preserving quotient systems in the sense that they preserve the closed properties of the original systems and can be used to reduce the complexity of verifying properties of very large systems. If a hybrid automaton has a finite state bisimulation, then checking properties for the hybrid automata can be equivalently performed on the finite, discrete, quotient graph. Since the quotient graph is finite, the algorithm will terminate. However, a finite state bisimulation exists only for certain hybrid automata classes, for which the reachability problem is decidable [161].

3.3 Randomness Issues when Modelling Hybrid Systems

For practical reasons, hybrid system modelling should allow uncertainty to enter in a number of places: choice of continuous evolution (modelled, for example, by a differential inclusion), choice of discrete transition destination or choice between continuous evolution and discrete transition. The concepts of stochastic and probabilistic hybrid systems are introduced to describe a large class of uncertain hybrid system models where uncertainty can be quantified probabilistically. The simplest probabilistic hybrid model considers probability distributions on the set of possible switches. Usually, the term '*probabilistic hybrid system*' is used when we deal only with discrete probability distributions. The need for finer probabilistic analysis of uncertain systems has led to the study of an even wider class of hybrid systems. This class allows for situations as random failures, which cause unexpected transitions from one discrete state to another, or random task execution times, which affect how long the system spends in different modes. In more complex form, plant behaviour can be affected by noise (stochastic disturbance) in one or more continuous components of the system, in which case we must resort to using stochastic differential equations. When the model is described by continuous probability distributions, we refer to it as a *stochastic hybrid system*.

For hybrid systems, there exist a bunch of techniques, methodologies and tools for design, control, analysis, simulation and verification. When we deal with randomness, these techniques/methodologies/tools have to be redesigned from scratch in order to consider probabilities. In most cases, the traditional approaches can no longer be used and new approaches should be discovered and employed.

In this section, we provide an overview of a number of stochastic hybrid processes (SHP) developed in the literature. Very well-studied stochastic processes, such as Levy processes and jump processes (see [83]) can be considered stochastic hybrid processes, even though the hybrid aspect of their dynamics is fairly weak. Closer to the framework developed in recent years for deterministic hybrid systems are piecewise deterministic Markov processes (PDMP) [64], switching diffusion processes (SDP) [105], stochastic hybrid automata (SHA) [122], stochastic hybrid models (SHM) [19] and the general switching diffusion processes (GSDP) [34]. Stochastic extensions of timed automata have been studied in [13, 151, 152].

All these stochastic hybrid processes can capture different classes of dynamics, depending on the applications they were developed to address. The most important difference among the models lies in where randomness is introduced. Some models allow diffusions to model continuous evolution [19, 34, 64, 122], while others do not [64, 83]. Likewise, some models force transitions to take place from certain states (e.g., [19, 122]), others only allow transitions to take place at random times (e.g., using a generalised Poisson process [105]), while others allow both [34, 64]. Here, we refer to the first type of transition as *forced transition* and to the latter as *spontaneous transition*.

The original formulation of many of these models includes control inputs [19, 34, 64, 103–105] but other models are autonomous [122]. To allow a uniform presentation we discuss only the autonomous versions for all models. All models are presented using a common formal language to facilitate the comparison; the notation used is based on the formalism developed for deterministic hybrid systems in [168]. A preliminary analysis of the models developed by Davis, Ghosh and Hu can be found in [193].

3.3.1 Probabilistic Hybrid Automata

Probabilistic hybrid models can be thought of as extensions of discrete models, such as hidden Markov models [82], Markov chains, semi-Markov processes [167] or dynamic Bayesian networks [57] to continuous dynamical models. In practice, the modelling of many phenomena requires the integration of both probabilistic and hybrid (mixed discrete continuous) aspects. Even though deterministic hybrid models can capture a wide range of behaviours encountered in practice, stochastic features are very important because of the uncertainty inherent in most real world applications. Compared to more traditional hybrid systems, probabilistic hybrid models have properties crucial to reasoning under uncertainty, including probabilistic transitions between modes or noisy observations.

In probabilistic hybrid automata (PHA) [118], a system is modelled by a hybrid automaton that has both discrete and continuous variables. This framework can be viewed as an extension of a hidden Markov model. The hidden Markov model is a finite set of states, each of which is associated with a probability distribution. Transitions among states are governed by a set of probabilities called transition probabilities. In a particular state, an outcome or observation can be generated according to the associated probability distribution. Only the outcome, not the state is visible to an external observer, and therefore states are hidden to the outside. A PHA incorporates discrete and continuous inputs, stochastic continuous dynamics and autonomous mode transitions. Simpler versions of PHA can be found in [216]. These are variants of hybrid automata augmented with discrete probability distributions.

Formally, a discrete-time PHA [118] is defined as a tuple

$$\mathscr{A} := \langle Q, \mathbf{x}_c, \mathbf{y}_c, \mathscr{U}, F_c, G_c, \mathscr{T} \rangle,$$

with the notation defined as below:

- The finite set Q denotes the modes $i \in Q$ of the automaton.
- $\mathbf{x}_c$ and $\mathbf{y}_c$ denotes the set of independent continuous state variables and output variables, respectively. The set of input variables

$$\mathscr{U} = \mathbf{u}_c \cup \mathbf{u}_d \cup \mathbf{v}_c$$

is divided into continuous control variables $\mathbf{u}_c$, continuous exogenous variables $\mathbf{v}_c$ and discrete control variables $\mathbf{u}_d$. Components of continuous variables range over different $\mathbb{R}^n$, whereas components of discrete variables range over finite domains D.

- The set F_c and G_c associate with each mode $i \in Q$ functions f_{ci} and g_{ci} that govern the continuous dynamics exhibited at mode i by (in terms of discrete-time difference equations and algebraic equations)

$$\mathbf{x}_c(k+1) = f_{ci}(\mathbf{x}_c(k), \mathbf{u}_c(k), \mathbf{v}_c(k)),$$
$$\mathbf{y}_c(k) = g_{ci}(\mathbf{x}_c(k), \mathbf{v}_c(k)).$$

- $\mathscr{T}$ specifies for each mode $i(k)$ (at the moment k) a set of transition functions $\mathscr{T}_i = \{\tau_{i1}, \ldots, \tau_{in}\}$. Each transition τ_{ij} has an associated guard condition $C_{ij}(x_c(k), u_d(k))$ and specifies the probability distribution over target modes $l(k+1)$ (at the moment $k+1$) together with an assignment for $x_c(k+1)$.

3.3.2 Piecewise Deterministic Markov Processes

Piecewise deterministic Markov processes (PDMP) were introduced by Davis in 1984 [65]. They are a class of non-linear continuous-time stochastic hybrid processes, which covers a wide range of nondiffusion phenomena. PDMPs involve a hybrid state space, i.e., with both continuous and discrete states. Randomness appears only in the discrete transitions; between two consecutive discrete transitions the continuous state evolves according to a non-linear ordinary differential equation. Discrete transitions occur either when the state hits the state space boundary or according to a generalised Poisson process in the interior of the state space. Whenever such a transition occurs, the hybrid state is reset instantly according to a probability distribution, which depends on the hybrid state before the transition.

Formally, we introduce PDMP following the notation of [42, 168, 193]. Let Q be a countable set of discrete states and let $d : Q \to \mathbb{N}$ and $\mathscr{X} : Q \to \mathbb{R}^{d(\cdot)}$ be two maps assigning to each discrete state $i \in Q$ an open subset of $\mathbb{R}^{d(i)}$. We call the set

$$\mathbb{S}(Q, d, X) := \bigcup_{i \in Q} \{i\} \times \mathscr{X}(i) = \{(i, z) : i \in Q, z \in \mathscr{X}(i)\}$$

the *PDMP hybrid state space* and

$$\mathbf{x} := (i, z) \in \mathbb{S}(Q, d, X)$$

the *hybrid state*. We define the boundary of the hybrid state space as

$$\partial \mathbb{S}(Q, d, X) := \bigcup_{i \in Q} \{i\} \times \partial \mathscr{X}(i).$$

A vector field f on the hybrid state space $\mathbb{S}(Q, d, X)$ is a function

$$f : \mathbb{S}(Q, d, X) \to \mathbb{R}^{d(\cdot)}$$

assigning to each hybrid state $\mathbf{x} = (i, z)$ a direction $f(\mathbf{x}) \in \mathbb{R}^{d(i)}$. The flow of f is a function

$$\phi : \mathbb{S}(Q, d, X) \times \mathbb{R} \to \mathbb{S}(Q, d, X)$$

with

$$\phi(\mathbf{x}, t) = \begin{bmatrix} \phi_Q(\mathbf{x}, t) \\ \phi_X(\mathbf{x}, t) \end{bmatrix},$$

where $\phi_Q(\mathbf{x}, t) \in Q$ and $\phi_X(\mathbf{x}, t) \in \mathscr{X}(i)$, such that for $\mathbf{x} = (i, z)$, $\phi(\mathbf{x}, 0) = \mathbf{x}$ and

$$\phi_Q(\mathbf{x}, t) = i,$$
$$\frac{d}{dt}\phi_X(\mathbf{x}, t) = f\big(\phi(\mathbf{x}, t)\big)$$

for all $t \in \mathbb{R}$.

Let

$$\Gamma\big((Q, d, \mathscr{X}), f\big) = \big\{\mathbf{x} \in \partial\mathbb{S}(Q, d, X) | \exists (\mathbf{x}', t) \in \mathbb{S}(Q, d, X) \times \mathbb{R}^+, \mathbf{x} = \phi(\mathbf{x}', t)\big\};$$

and

$$\overline{\mathbb{S}}(Q, d, \mathscr{X}) = \mathbb{S}(Q, d, X) \cup \Gamma\big((Q, d, \mathscr{X}), f\big).$$

Define

$$\mathscr{B}(\overline{\mathbb{S}}) := \sigma\left(\bigcup_{i \in Q}\{i\} \times \mathscr{B}(i)\right),$$

where $\overline{\mathbb{S}} = Q \times \mathbb{R}^{\infty}$. The space $(\overline{\mathbb{S}}, \mathscr{B}(\overline{\mathbb{S}}))$ is a Borel space and $\mathscr{B}(\overline{\mathbb{S}})$ is a sub-σ-algebra of its Borel σ-algebra. We can now introduce the following definition.

Definition 3.5 (Piecewise deterministic Markov process) A piecewise deterministic Markov process is a collection

$$H = \big((Q, d, \mathscr{X}), f, \text{Init}, \lambda, R\big)$$

where:

- Q is a finite/countable set of discrete variables;
- $d : Q \to \mathbb{N}$ is a map giving the dimensions of the continuous state spaces;
- $\mathscr{X} : Q \to \mathbb{R}^{d(\cdot)}$ maps each $i \in Q$ into a subset $\mathscr{X}(i)$ of $\mathbb{R}^{d(i)}$;
- $f : \mathbb{S}(Q, d, X) \to \mathbb{R}^{d(\cdot)}$ is a vector field;
- Init $: \mathscr{B}(\overline{\mathbb{S}}) \to [0, 1]$ is an initial probability measure on $(\overline{\mathbb{S}}, \mathscr{B}(\overline{\mathbb{S}}))$, with $\text{Init}(\mathbb{S}^c) = 0$;
- $\lambda : \overline{\mathbb{S}}(Q, d, \mathscr{X}) \to \mathbb{R}^+$ is a transition rate function;
- $R : \mathscr{B}(\overline{\mathbb{S}}) \times \overline{\mathbb{S}}(Q, d, \mathscr{X}) \to [0, 1]$ is a transition measure, with $R(\mathbb{S}^c, \cdot) = 0$.

To ensure the process is well defined, the following assumption is introduced in [65].

Assumption 3.1

(i) The sets $\mathscr{X}(i)$ are open, for all $i \in Q$.
(ii) For all $i \in Q$, $f(i, .)$ is globally Lipschitz continuous.
(iii) The transition rate function

$$\lambda : \overline{\mathbb{S}}(Q, d, \mathscr{X}) \to \mathbb{R}^+$$

is measurable.
(iv) For all $x \in \mathbb{S}$, there exists $\varepsilon > 0$ such that the function $t \to \lambda(\phi(\mathbf{x}, t))$ is integrable for all $t \in [0, \varepsilon)$.
(v) For all $A \in \mathscr{B}(\overline{\mathbb{S}})$, $R(A, \cdot)$ is measurable.

To define the PDMP executions we need to introduce the notion of exit time $t^* : \mathbb{S} \to \mathbb{R}^+ \cup \{\infty\}$ as

$$t^*(\mathbf{x}) = \inf\big\{t > 0 : \phi(\mathbf{x}, t) \notin \mathbb{S}\big\}$$

and of survivor function $F : \mathbb{S} \times \mathbb{R}^+ \to [0, 1]$ as

$$F(\mathbf{x}, t) = \begin{cases} \exp(-\int_0^t \lambda(\phi(\mathbf{x}, \tau)) \, d\tau) & \text{if } t < t^*(\mathbf{x}), \\ 0 & \text{if } t \geq t^*(\mathbf{x}). \end{cases}$$

The executions of the PDMP can be thought of as being generated by the following algorithm (see Fig. 3.3).

Algorithm 3.1 (Generation of PDMP executions)
```
set T = 0
select S-valued random variable α̂ according to Init
repeat
    select R⁺-valued random variable T̂
```

Fig. 3.3 Executions of PDMPs

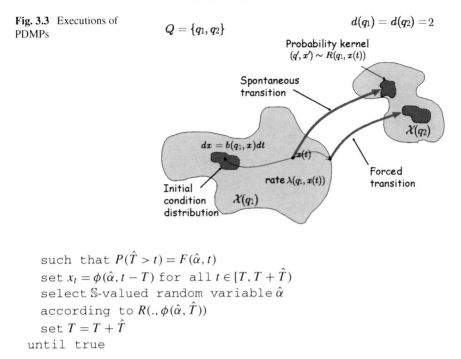

$$Q = \{q_1, q_2\}$$

$$d(q_1) = d(q_2) = 2$$

Probability kernel
$(q', x') \sim R(q_1, x(t))$

Spontaneous transition

$\mathcal{X}(q_2)$

$dx = b(q_1, x)dt$

$x(t)$

Forced transition

Initial condition distribution

rate $\lambda(q_1, x(t))$

$\mathcal{X}(q_1)$

```
such that P(T̂ > t) = F(α̂, t)
set x_t = φ(α̂, t − T) for all t ∈ [T, T + T̂)
select S-valued random variable α̂
according to R(., φ(α̂, T̂))
set T = T + T̂
until true
```

All random extractions in Algorithm 3.1 are assumed to be independent. To ensure that x_t is defined on the entire $\mathbb{R}^+$ it is necessary to exclude Zeno executions [168]. The following assumption is introduced in [64] to accomplish this.

Assumption 3.2 Let

$$N_t = \sum_i I_{(t \geq T_i)}$$

be the number of jumps in $[0, t]$. Then $\mathbb{E}[N_t] < \infty$ for all t.

Under Assumptions 3.1 and 3.2, Algorithm 3.1 defines a strong Markov process [64, 65].

3.3.3 Stochastic Hybrid Systems

Stochastic hybrid automata (SHA) introduced in [122] are another class of non-linear, continuous-time stochastic hybrid processes. SHA also involve a hybrid state space, with both continuous and discrete states. The continuous state evolves according to a stochastic differential equation that depends on the discrete state. Transitions occur when the continuous state hits the boundary of the state space (forced transitions). Whenever a transition occurs the hybrid state is reset instantaneously to a new value. The value of the discrete state after the transition is deterministically given by the hybrid state before the transition. The new value of the continuous state, on the other hand, is governed by a probability law which depends on the last hybrid state.

Definition 3.6 (Stochastic hybrid automaton) A stochastic hybrid automaton is a collection

$$H = (Q, X, \text{Dom}, f, g, \text{Init}, G, R)$$

where

- Q is a countable set representing the discrete state space;
- $X = \mathbb{R}^n$ is the continuous state space;
- $\text{Dom} : Q \to 2^X$ assigns to each $i \in Q$ an open subset of X;
- $f, g : Q \times X \to \mathbb{R}^n$ are vector fields;
- $\text{Init} : \mathscr{B}(Q \times X) \to [0, 1]$ is an initial probability measure on $(Q \times X, \mathscr{B}(Q \times X))$ concentrated on $\bigcup_{i \in Q} \{i\} \times \text{Dom}(i)$;
- $G : Q \times Q \to 2^X$ assigns to each $(i, j) \in Q \times Q$ a guard $G(i, j) \subset X$ such that

  (a) for each $(i, j) \in Q \times Q$, $G(i, j)$ is a measurable subset of $\partial\text{Dom}(i)$ (possibly empty);
  (b) for each $i \in Q$, the family $\{G(i, j) \mid j \in Q\}$ is a disjoint partition of $\partial\text{Dom}(i)$;

- $R : Q \times Q \times X \to \mathscr{P}(X)$ assigns to each $(i, j) \in Q \times Q$ and $x \in G(i, j)$ a reset probability kernel on X concentrated on $\text{Dom}(j)$.

We again use $\alpha = (q, x)$ to denote the hybrid state of an SHA. To ensure that the model is well defined the following assumption is introduced in [122].

Assumption 3.3

(i) For all $i \in Q$ the functions $f(i, \cdot)$ and $g(i, \cdot)$ are bounded and Lipschitz continuous.
(ii) For all $i, j \in Q$ and for any measurable set $A \subset \text{Dom}(j)$, $R(i, j, x)(A)$ is a measurable function in x.

The first part of Assumption 3.3 ensures that for any $i \in Q$, the solution of the SDE:

$$dx(t) = f\big(i, x(t)\big) dt + g\big(i, x(t)\big) dW_t,$$

where W_t is a 1-dimensional standard Wiener process, exists and is unique (see Theorem 6.2.2 in [7]). Moreover, the assumption on R ensures that 'transition' events are measurable with respect to the underlying σ-field, hence their probabilities make sense.

Fig. 3.4 Executions of SHSs

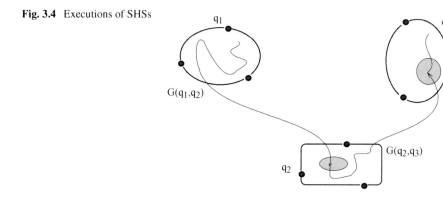

Definition 3.7 (SHA execution) A stochastic process $x_t = (q(t), z(t))$ is called an SHA execution (see Fig. 3.4) if there exists a sequence of stopping times

$$T_0 = 0 \le T_1 \le T_2 \le \cdots$$

such that for each $j \in \mathbb{N}$,

- $\mathbf{x}(0) = (q(0), z(0))$ is a $(Q \times X)$-valued random variable with distribution Init;
- for $t \in [T_j, T_{j+1})$, $q(t) = q(T_j)$ is constant and $z(t)$ is a (continuous) solution of the SDE:

$$dz(t) = f\big(q(T_j), z(t)\big) dt + g\big(q(T_j), z(t)\big) dW_t$$

starting at $z(T_j)$, where W_t is a 1-dimensional standard Wiener process;
- $T_{j+1} = \inf\{t \ge T_j | z(t) \notin \mathrm{Dom}(q(T_j))\}$;
- $\lim_{t \uparrow T_{j+1}} z(t) \in G(q(T_j), q(T_{j+1})) \in \partial\mathrm{Dom}(q(T_j))$;
- $z(T_{j+1})$ is a random variable distributed according to $R(q(T_j), q(T_{j+1}),$ $\lim_{t \uparrow T_{j+1}} z(t))$.

In [122], it is shown that under Assumption 3.3 $\{\mathbf{x}_{T_n}\}$ defines a Markov process.

3.3.4 Switching Diffusion Processes

Switching diffusion processes (SDPs) were introduced by Ghosh et al. in 1991. They are a class of non-linear, continuous-time SHPs that have been used to model a large number of applications such as fault tolerant control systems, multiple target tracking and flexible manufacturing systems. SDPs exhibit also a sort of hybrid dynamics. The continuous state evolves according to a stochastic differential equation (SDE), while the discrete state is a controlled Markov chain. Both the dynamics of the SDE and the transition matrix of the Markov chain depend on the hybrid state. The continuous hybrid state evolves without jumps, i.e., the evolution of the continuous state can be assumed to be a continuous function of time.

In the following we introduce formally SDP following [103–105]. To allow a comparison with other SHP we restrict our attention to autonomous SDP.

Definition 3.8 (Switching diffusion process) A switching diffusion process is a collection

$$H = (Q, X, f, \text{Init}, \sigma, \lambda_{ij})$$

where:

- $Q = \{1, 2, \ldots, N\}$ is a finite set of discrete variables, $N \in \mathbb{N}$;
- $X = \mathbb{R}^n$ is the continuous state space;
- $f : Q \times X \to \mathbb{R}^n$ is a vector field;
- $\text{Init} : \mathcal{B}(Q \times X) \to [0, 1]$ is an initial probability measure on $(Q \times X, \mathcal{B}(Q \times X))$;
- $\sigma : Q \times X \to \mathbb{R}^{n \times n}$ is a state-dependent real-valued matrix;
- $\lambda_{ij} : X \to \mathbb{R}, i, j \in Q$ are a set of z-dependent transition rates, with

$$\lambda_{ij}(.) \geq 0 \quad \text{if } i \neq j; \quad \sum_{j \in Q} \lambda_{ij}(z) = 0$$

for all $i \in Q, z \in X$.

We denote with

$$\mathbf{x} := (q, z)$$

the hybrid state of an SDP. To ensure the SDP model is well defined [103–105], we introduce the following assumption.

Assumption 3.4 The functions $f(i, z)$, $\sigma_{kj}(i, z)$ and $\lambda_{kj}(z)$ are bounded and Lipschitz continuous in z.

Assumption 3.4 ensures that for any $i \in Q$, the solution to the SDE

$$dz(t) = f\big(i, z(t)\big) dt + \sigma\big(i, z(t)\big) dW_t,$$

where W_t is an n-dimensional standard Wiener process, exists and is unique.

For $i, j \in Q$ and $z \in \mathbb{R}^n$ let $\Delta(i, j, z)$ be consecutive, with respect to lexicographic ordering on $Q \times Q$, left closed, right open intervals of the real line, each having length $\lambda_{ij}(z)$ (for details see [103–105]). Now define a function

$$h : \mathbb{R}^n \times Q \times \mathbb{R} \to \mathbb{R}$$

by setting

$$h(z, i, y) := \begin{cases} j - i, & \text{if } z \in \Delta(i, j, z), \\ 0, & \text{otherwise.} \end{cases} \tag{3.1}$$

SDP executions can be defined using h.

Fig. 3.5 Executions of SDPs

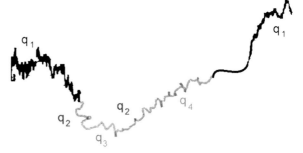

Definition 3.9 (SDP execution, see Fig. 3.5) A stochastic process $x_t = (q(t), z(t))$ is called an SDP execution if it is the solution of the following SDE and stochastic integral:

$$dz(t) = f\big(q(t), z(t)\big)\, dt + \sigma\big(q(t), z(t)\big)\, dW_t,$$

$$dq(t) = \int_{\mathbb{R}} h\big(u(t), q\big(t^-\big), z\big)\varphi(dt, du)$$

for $t \geq 0$ with $z(0) = z_0$, $q(0) = q_0$, where

- $x_0 = (q(0), z(0))$ is a random variable extracted according to the probability measure Init;
- W_t is an n-dimensional standard Wiener process;
- $\varphi(dt, du)$ is an $\mathcal{M}(\mathbb{R}^+ \times \mathbb{R})$-valued Poisson random measure with intensity $dt \times m(dz)$, where m is the Lebesgue measure on $\mathbb{R}$ (see [137]);
- $\varphi(.,.)$, W_t and $(q(0), z(0))$ are independent.

3.3.5 General Switching Diffusion Processes

General switching diffusion processes (GSDP) were introduced by Borkar et al. in 1999 [34]. They are a class of non-linear, continuous-time, controlled SHP. To allow a comparison with the stochastic hybrid processes previously described, we restrict our attention to autonomous GSDP. GSDP involve also a hybrid state space. This model allows two types of transition: spontaneous transitions and forced transitions. The discrete state evolves as a Markov chain whose transition rate depends on the hybrid state before the jump and it coincides with the SDP transitions: these are the spontaneous transitions. The hybrid state switches from one diffusion path to another during the spontaneous transitions in a continuous way, without resetting (changing) the continuous state. Forced transitions occur when the continuous state hits a given subset of the hybrid state space. When a forced transition occurs, the continuous state is reset according to a probability distribution that depends on the hybrid state before the transition. Finally, in the latter type of transition, a time delay is introduced to allow noninstantaneous transitions.

In this section, we give a formal model of GSDP following [34], without considering delayed transitions, since they are absent in all the previously discussed stochastic hybrid models. However, delayed transitions could be introduced by addition of more discrete states to the model.

Definition 3.10 (General switching diffusion process) A general switching diffusion process is a collection

$$H = (Q, d, X, f, \text{Init}, \sigma, \lambda_{ij}, A, D, R)$$

where

- $Q = \{1, 2, \ldots, N\}$ is a finite set of discrete variables, $N \in \mathbb{N}$;
- $d : Q \to \mathbb{N}$ is a map giving the dimension of the continuous state space;
- $X = (\bigcup_{n=1}^{\infty} X_n)$ is the continuous state space, where X_n is the closure of a connected open subset of some Euclidean space $\mathbb{R}^{d(n)}$;
- $f_n : Q \times X_n \to \mathbb{R}^{d(n)}$ is a vector field;
- $\text{Init} : \mathscr{B}(Q \times X) \to [0, 1]$ is an initial probability measure on $(Q \times X, \mathscr{B}(Q \times X))$;
- $\sigma_n : Q \times X_n \to \mathbb{R}^{d(n) \times d(n)}$ is a state-dependent real-valued matrix;
- $\lambda_{ij} : X \to \mathbb{R}$, $i, j \in Q$ are a set of z-dependent transition rates, with $\lambda_{ij}(\cdot) \geq 0$ if $i \neq j$ and $\sum_{j \in Q} \lambda_{ij}(z) = 0$ for all $i \in Q$, $z \in X$.
- $A = (\bigcup_{n=1}^{\infty} A_n)$ is the set-interface, where $A_n \subset X_n$ closed;
- $D = (\bigcup_{n=1}^{\infty} D_n)$ is the destination set, where $D_n \subset X_n$ closed and $A_n \cap D_n = \varnothing$ for any $n \in \mathbb{N}$;
- $R : A \times Q \to \mathscr{P}(D)$ assigns a conditional law of jump destination for forced jumps.

We denote with $\mathbf{x} = (q, z)$ the hybrid state of a GSDP. To ensure the GSDP model is well defined, the following assumption is introduced in [34].

Assumption 3.5

(i) The functions $f^n(q, z)$, $\sigma_{kj}^n(q, z)$ and $\lambda_{kj}(z)$ are bounded and Lipschitz continuous in z.

(ii) The least eigenvalue of $\sigma_{kj}^n(q, z)(\sigma_{kj}^n(q, z))^T$ is uniformly bounded away from zero.

(iii) The map R is bounded and uniformly continuous.

(iv) ∂A_n is C^2 and let $\partial S_n \subset \partial A_n$; furthermore, ∂A_n is a C^2-manifold without boundary; for each n, $\inf_n d(A_n, D_n) > 0$.

(v) Finally, each D_n is a bounded set.

Thanks to the first part of Assumption 3.5, it is possible to ensure that for any $q \in Q$, the solution to the SDE

$$dx(t) = f_n\big(q, z(t)\big)\, dt + \sigma_n\big(q, z(t)\big)\, dW_t^n,$$

where W_t^n is a $d(n)$-dimensional standard Wiener process, exists and is unique.

Definition 3.11 (GSDP execution) A stochastic process $x_t = (q(t), z(t))$ is called a GSDP execution if there exists a sequence of stopping times $T_0 = 0 \le T_1 \le T_2 \le \cdots$ such that for each $j \in \mathbb{N}$,

- $x_0 = (q(0), z(0))$ is a $(Q \times X)$-valued random variable extracted according to the probability measure Init;
- $z(t)$ and $q(t)$ are the solution of the following SDE and stochastic integral:

$$dz(t) = f_n\big(q(t), z(t)\big)\, dt + \sigma_n\big(q(t), z(t)\big)\, dW_t^n,$$

$$dq(t) = \int_{\mathbb{R}} h\big(u(t), q(t^-), u\big)\varphi(dt, du),$$

where h is a suitable function defined like in the SDP case (see (3.1)); W_t^n is a $d(n)$-dimensional standard Wiener process; $\varphi(dt, du)$ is an $\mathcal{M}(\mathbb{R}^+ \times \mathbb{R})$-valued Poisson random measure with intensity $dt \times m(dz)$, where m is the Lebesgue measure on $\mathbb{R}$ [137]; $\varphi(\cdot, \cdot)$, W_t and $(q(0), z(0))$ are independent;

- $T_{j+1}^- = \inf\{t \ge T_j : z(t) \in A\}$; $z(T_{j+1}^-) \in A$;
- the probability distribution of $z(T_{j+1}^-) \in D$ is governed by the law

$$R\big(z(T_{j+1}^-), q(T_{j+1}^-)\big).$$

The above description sets up GSDPs as a very general class of stochastic hybrid processes. Indeed, SDPs and SHAs can be considered special subclasses of it. On the contrary, PDMPs cannot be viewed as a special subclass of GSDP because the latter do not allow spontaneous reset of the continuous hybrid state. Unfortunately, in [34], there is no proof that GSDPs are Markov processes and there are no characterisations of the operator semigroup or infinitesimal generator, etc. Then a formal comparison with the previously described SHP is not possible.

Another model for stochastic hybrid processes is given by switching jump-diffusions, developed in [26]. Switching jump-diffusion processes (SJDP) have jumps that:

(i) happen simultaneously with mode switching, and
(ii) depend on the mode after switching.

Jumps satisfying both (i) and (ii) are referred to as *hybrid jumps*. There are two types of hybrid jump: a forced jump, which happens at an instant of hitting some boundary and a Poisson type of jump, which happens at a sudden instant. It is important to point out that the term hybrid jump was introduced for the first time in [26] and the mathematical basis for it was developed in [27].

3.3.6 Analysis of Stochastic Hybrid Systems

One research direction in the analysis of stochastic hybrid systems is the extension of the techniques developed previously for Markov chains to include systems with simple continuous dynamics. This approach, developed mostly in computer science, extends *probabilistic model checking*, developed in early research for Markov chains [9].

Only a few model-checking approaches to verify stochastic hybrid systems exist. For example, Kwiatkowska et al. [150] discuss a probabilistic symbolic model-checking software called PRISM. PRISM is used for automatic verification of probabilistic computation tree logic (PCTL) properties.

One very successful verification method for hybrid systems is *bounded model checking* (BMC) [24, 92]. BMC was originally formulated for discrete transition systems and then extended to hybrid systems. Given a horizon of interest, BMC verifies whether all finite-horizon trajectories satisfy a temporal logic formula into two steps. It

1. translates the problem into a large satisfiability SAT-problem [213], and
2. finds a counterexample, based on extremely powerful state-of-the-art SAT-solvers.

BMC encodes the next-state relation of a system as a propositional formula, then unrolls this to some given finite depth k, augmented with a corresponding finite unravelling of the tableau of (the negation of) a temporal formula. The goal is to obtain a propositional SAT problem, which is satisfiable if and only if an error trace of length k exists. The BMC formulae arising from such systems comprise complex Boolean combinations of arithmetic constraints over real-valued variables, thus necessitating satisfiability-modulo-theory (SMT) solvers over arithmetic theories to solve them [91]. In [91], a stochastic version of SMT-based BMC has been developed for reasoning about simple probabilistic hybrid models and their requirements.

This probabilistic model checker can deal only with simple models of probabilistic hybrid automata where the continuous evolution is deterministic (usually given by linear differential equations) and the discrete transitions are governed by discrete probability distributions.

One way to apply existing probabilistic model checkers to stochastic hybrid systems is by Markov chain approximations [4]. The key idea is the use of numerical tools to generate a finite state Markov chain from a given stochastic hybrid system, which comes with a certain level of approximation error. The properties of the Markov chain (the probability to evolve in a certain region of the state space) are then analysed by a probabilistic model checker. Then we may obtain an overall guarantee about the probability of satisfying the original property of interest in the given stochastic hybrid system by combining the model-checker results with the error guarantee of the approximation scheme. This approach has been applied only to stochastic hybrid systems with discrete time. The only model-checking approach that exists for continuous-time stochastic hybrid systems uses PDMP as the underlying model and it has been reported in [49].

Another research direction in the analysis of stochastic hybrid systems is extension of techniques developed previously for continuous state space and continuous-time stochastic dynamical systems and control. For the analysis of such systems, because of the stochastic specification of stochastic hybrid systems, one can employ probabilistic methods, as well as analytic methods of optimal control.

A computational method that characterises reachability and safety as a viscosity solution of a system of coupled Hamilton–Jacobi–Bellman equations analyses stochastic hybrid systems by computing a solution based on discrete approximations [146].

Probabilistic reachability analysis techniques based on dynamic programming have been developed for controlled discrete-time stochastic hybrid systems [2] and for large-scale stochastic hybrid systems using rare event estimation theory [29].

Simulation techniques to analyse stochastic hybrid system evolutions have proven quite useful to biological systems modelling [205]. For verification purposes, when the state space of stochastic hybrid systems becomes too large to let us use model-checking and reachability analysis techniques, Markov chain Monte Carlo techniques can be used to approximate a likelihood of system failure [148].

At the end of this section, we outline the fact that analysis methods for stochastic hybrid systems are built mostly on their probabilistic characterisations. Formal or control methods provide results in terms of probabilities, expectations, moments, distributions, etc. The solutions to practical problems depend on our capacity to handle and interpret such results.

3.4 Some Remarks

This chapter offered highlights on modelling hybrid systems from deterministic to stochastic. This presentation is not meant to be exhaustive, but rather summarises the major results regarding the above topics, and other results that are particularly relevant to this monograph. For deterministic hybrid systems, the modelling and analysis approaches abound. We have provided only the references for the 'classical' approaches, but many others are also important in the development of this theory.

Chapter 4
Stochastic Hybrid Systems

4.1 Overview

In the face of the growing complexity of control systems, stochastic modelling has taken on a crucial role. Indeed, stochastic techniques for modelling control and hybrid systems have attracted the attention of many researchers and constitute one of the most important issues of contemporary research.

Hybrid systems have been extensively studied in the past decade, with regard to both the theoretical framework and the increasing number of applications, for which they are employed. However, the subfield of stochastic hybrid systems is fairly young. There has been considerable current interest in stochastic hybrid systems due to their ability to represent such systems as manoeuvring aircraft [132] and switching communication networks [117]. Other areas to which stochastic hybrid systems have found applications include insurance pricing [63], capacity expansion models for the power industry [66], flexible manufacturing and fault tolerant control [104, 105].

A considerable amount of research has been directed towards this topic, both in the direction of extending the theory of deterministic hybrid systems [122] and the discovery of new applications unique to the probabilistic framework.

Objectives   In this chapter, we tackle the following issues:

1. Introduction of a very general model for stochastic hybrid processes: the general stochastic hybrid system (GSHS).
2. Development of a mathematical construction for piecing out Markov processes that preserves the Markov property.
3. Show how the behaviour of a GSHS can be described as piecing out diffusion processes and then deriving the basic mathematical properties of a GSHS.

A GSHS can be thought of as a 'traditional' hybrid system where discrete and continuous behaviours are randomised in the following manner:

1. Continuous-time dynamics are driven by stochastic differential equations rather than classical ordinary differential equations.

L.M. Bujorianu, *Stochastic Reachability Analysis of Hybrid Systems*,
Communications and Control Engineering,
DOI 10.1007/978-1-4471-2795-6_4, © Springer-Verlag London Limited 2012

2. Discrete transitions (jumps) take place when the continuous state hits a 'guard' set (that could be the mode boundary, or an interface set), or they occur according to a transition rate.
3. Postjump locations are randomly chosen according to probability that depends also on the prejump state (a stochastic kernel).

The GSHS realisation can be described as an interleaving of a finite or countable family of diffusion processes and a jump process. The scope of this chapter is to prove that the GSHS is indeed a sound probabilistic model that provides the suitable framework necessary to investigate inherent properties of control systems that we intend to model. A natural property we sought was the Markov property. The first analysis of the form of GSHS executions (paths or trajectories) shows that these are, in fact, 'concatenations' of the component diffusion paths. The continuity inherited from the diffusion trajectories is perturbed only by the discrete transitions.

This observation leads to the investigation of a general construction for piecing out Markov processes that preserves the Markov property. Given a finite or countable family of Markov processes with reasonably good properties, this mechanism will provide a method to obtain new interesting Markov processes whose paths are obtained by 'sticking' together the component paths. The result of this mixing operation will be called a *Markov string*. Roughly speaking, Markov strings are composites of Markov processes. The jump structure of a Markov string is completely described by a renewal kernel given a priori and a family of stopping times associated with the initial processes. One major assumption is that Markov strings are allowed to have only finitely many jumps in finite time. Under these assumptions we prove that the Markov strings, as stochastic processes, have the strong Markov property and, componentwise, the trajectories inherit the continuity properties of the components.

At the end of the chapter, we show how a GSHS can be embedded in the modelling framework of Markov strings. Then the behaviour of a GSHS inherits the strong Markov property from Markov strings, and the GSHS trajectories are right continuous with left limits. We must stress that the strong Markov property is a fundamental hypothesis of stochastic reachability. Other properties regarding the transition probabilities, operator semigroup and infinitesimal generator are also discussed. Finally, the mathematical expression of the infinitesimal generator associated to a GSHS is given. In addition, we discuss how the infinitesimal generator characterises the hybrid behaviour of GSHS.

4.2 General Stochastic Hybrid Systems

This section presents the most general stochastic hybrid automaton model, which can be obtained by randomisation of the classical deterministic hybrid automaton. This dynamics randomisation acts on both discrete and continuous components. When one studies the behaviour of such systems, some knowledge of stochastic processes is compulsory. To facilitate an optimal understanding of the mathematical objects used in this chapter, this section and the following one start with some informal presentations.

4.2.1 Informal Presentation

General stochastic hybrid systems represent a class of nonlinear stochastic continuous-time hybrid dynamical systems. They are characterised by a hybrid state and hybrid dynamics. The hybrid state is defined by two components: the continuous state and the discrete state. The continuous and the discrete parts of the state variable have their own natural dynamics, governed by specific equations and reset maps, respectively. The most important feature of GSHS specification is how it captures the interaction between the discrete and continuous dynamics.

Time $t \geq 0$ is a continuous parameter. The state of the system is described by a continuous variable x and a discrete variable q. The continuous variable evolves in some modes X^q (open sets in the Euclidean space) and the discrete variable belongs to a countable set Q. The intrinsic difference between the discrete and continuous variables is apparent in the way they change over time. The continuous state evolves according to an SDE whose two parameters, vector field and drift factor, depend on the hybrid state. The discrete dynamics produce transitions in both continuous and discrete state variables x and q. Switching between two discrete states is governed by a probability law or it occurs when the continuous state hits the boundary of its mode. Whenever a switching occurs, the hybrid state is reset instantly to a new state according to a probability law, which itself depends on the past hybrid state. Transitions that occur when the continuous state hits the boundary of the state space are called *forced transitions*, and those which occur probabilistically according to a state-dependent rate are called *spontaneous transitions*. Thus, a sample trajectory has the form $(q_t, x_t, t \geq 0)$, where $(x_t, t \geq 0)$ is piecewise continuous and $q_t \in Q$ is piecewise constant. We let

$$0 \leq T_1 < T_2 < \cdots < T_i < T_{i+1} < \cdots$$

be the sequence of jump times.

It is easy to show many classes of stochastic hybrid processes found in the literature PDMPs, SDPs, GSDPs, SHSMs are particular instantiations of GSHSs. A GSHS generalises a PDMP, allowing a stochastic evolution (diffusion process) between two consecutive jumps, whilst for PDMP the inter-jump motion is deterministic, according to a vector field. In addition, a GSHS might be thought of as an extended SHA, for which the transitions between modes are triggered by some stochastic event (boundary hitting time or transition rate). Moreover, GSHSs generalise SDPs also, permitting the continuous state to have discontinuities when the process jumps from one diffusion to another. The following table shows how the GSHSs generalise different models of stochastic hybrid systems.

	PDMP	SHA	SDP	SJDP	GSHS
Stochastic continuous evolution	non	✓	✓	✓	✓
Forced transitions	✓	✓	non	✓	✓
Spontaneous transitions	✓	non	✓	✓	✓

For a detailed comparison of the first three models, the reader is referred to [193].

It can be shown also that the class of SJDP models can be considered a subclass of GSHSs whose stochastic kernel, which gives the postjump locations, is chosen in an appropriate way such that the change of the discrete state and continuous state at a jump depends on the postjump location (continuous and discrete).

In the remaining part of this chapter, we make use of some standard notions from Markov process theory: underlying probability space, natural filtration, translation operator, Wiener probabilities, admissible filtration, stopping times and strong Markov property [33].

4.2.2 The Mathematical Model

If X is a Hausdorff topological space we use to denote by $\mathscr{B}(X)$ or $\mathscr{B}$ its Borel σ-algebra (the σ-algebra generated by all open sets).

State Space   Let Q be a countable/finite set of discrete states, and let $d : Q \to \mathbb{N}$ and $\mathscr{X} : Q \to \mathbb{R}^{d(.)}$ be two maps assigning to each discrete state $q \in Q$ an open subset X^q of $\mathbb{R}^{d(q)}$. We call the set

$$\mathbf{X}(Q, d, \mathscr{X}) = \bigcup_{q \in Q} \{q\} \times X^q$$

the hybrid state space of the GSHS and we call

$$\mathbf{x} = (q, x^q) \in \mathbf{X}(Q, d, \mathscr{X})$$

the hybrid state. The closure of the hybrid state space will be

$$\overline{\mathbf{X}} = \mathbf{X} \cup \partial \mathbf{X},$$

where the boundary of $\mathbf{X}$ is

$$\partial \mathbf{X} = \bigcup_{q \in Q} \{q\} \times \partial X^q.$$

It is clear that, for each $q \in Q$, the state space X^q is a Borel space. One can define a metric ρ on X such that $\rho(\mathbf{x}_n, \mathbf{x}) \to 0$ as $n \to \infty$ with $\mathbf{x}_n = (q_n, x_n^{q_n})$, $\mathbf{x} = (q, x^q)$ if and only if there exists m such that $q_n = q$ for all $n \geq m$ and $x_{m+k}^q \to x^q$ as $k \to \infty$. The metric ρ restricted to any component X^q is equivalent to the usual Euclidean metric [64]. Each $\{q\} \times X^q$, being a Borel space, will be homeomorphic to a measurable subset of the Hilbert cube, $\mathscr{H}$ (Urysohn's theorem, Proposition 7.2

[22]). It is known that $\mathcal{H}$ is the product of countably many copies of $[0, 1]$. The definition of $\mathbf{X}$ shows that it is homeomorphic to a measurable subset of $\mathcal{H}$. Then $(X, \mathcal{B}(X))$ is a Borel space. Moreover, X is a Lusin space because it is a locally compact Hausdorff space with countable base (see [64] and the references therein).

Continuous and Discrete Dynamics   In each mode X^q, continuous evolution is driven by the SDE

$$dx_t^q = b\left(q, x_t^q\right) dt + \sigma\left(q, x_t^q\right) dW_t, \tag{4.1}$$

where $(W_t, t \geq 0)$ is the m-dimensional standard Wiener process in a complete probability space. The discrete component remains constant, i.e.,

$$q_t = q.$$

Assumption 4.1 (Continuous evolution) Suppose that

$$b : Q \times X^{(\cdot)} \to \mathbb{R}^{d(\cdot)},$$
$$\sigma : Q \times X^{(\cdot)} \to \mathbb{R}^{d(\cdot) \times m}, \quad m \in \mathbb{N},$$

are bounded and Lipschitz continuous in x.

This assumption ensures for any $q \in Q$ the existence and uniqueness (Theorem 6.2.2 in [7]) of the solution to the above SDE.

In this way, when q runs in Q, Eq. (4.1) defines a family of diffusion processes

$$M^q = \left(\Omega^q, \mathcal{F}^q, \mathcal{F}_t^q, x_t^q, \theta_t^q, \mathbb{P}^q\right), \quad q \in Q$$

with state spaces $\mathbb{R}^{d(q)}$, $q \in Q$. For each $q \in Q$, the elements $\mathcal{F}^q, \mathcal{F}_t^q, \theta_t^q, \mathbb{P}^q, \mathbb{P}_{x^q}^q$ have the usual meaning, as they have in the Markov process theory.

The jump (switching) mechanism between the diffusions is governed by two functions: the *jump rate* λ and the *reset kernel* R. The jump rate

$$\lambda : \mathbf{X} \to \mathbb{R}_+$$

is a measurable bounded function. The transition measure R maps $\overline{\mathbf{X}}$ into the set $\mathcal{P}(\mathbf{X})$ of probability measures on $(\mathbf{X}, \mathcal{B}(\mathbf{X}))$. Alternatively, one can consider the transition measure

$$R : \overline{\mathbf{X}} \times \mathcal{B}(\mathbf{X}) \to [0, 1]$$

as a reset probability kernel.

Assumption 4.2 (Discrete transitions)

 (i) For all $A \in \mathcal{B}(\mathbf{X})$, $R(\cdot, A)$ is measurable.
 (ii) For all $\mathbf{x} \in \overline{\mathbf{X}}$ the function $R(\mathbf{x}, \cdot)$ is a probability measure.
 (iii) $\lambda : \mathbf{X} \to \mathbb{R}_+$ is a measurable function such that

$$t \to \lambda\left(q, x_t^q\left(\omega^q\right)\right)$$

is integrable on $[0, \varepsilon(\omega^q))$ for some $\varepsilon(\omega^q) > 0$, for each $\omega^q \in \Omega^q$.

Since $\overline{\mathbf{X}}$ is a Borel space, $\overline{\mathbf{X}}$ is homeomorphic to a subset of the Hilbert cube $\mathscr{H}$. Therefore, its space of probabilities is homeomorphic to the space of probabilities of the corresponding subset of $\mathscr{H}$ (Lemma 7.10 [22]). There exists a measurable function $F : \mathscr{H} \times \overline{\mathbf{X}} \to \mathbf{X}$ such that

$$R(\mathbf{x}, A) = \mathfrak{p}\big[F^{-1}(A)\big], \quad A \in \mathscr{B}(X),$$

where $\mathfrak{p}$ is the probability measure on $\mathscr{H}$ associated to $R(\mathbf{x}, \cdot)$ and

$$F^{-1}(A) = \big\{\boldsymbol{\omega} \in \mathscr{H} \big| F(\boldsymbol{\omega}, \mathbf{x}) \in A\big\}.$$

The measurability of such a function is guaranteed by the measurability properties of the transition measure R.

Construction   We construct a GSHS as a *Markov 'sequence'* H, which admits (M^q) as subprocesses. A sample path of the stochastic process $(\mathbf{x}_t)_{t>0}$ with values in $\mathbf{X}$, starting from a fixed initial point $\mathbf{x}_0 = (q_0, x_0^{q_0}) \in \mathbf{X}$ is defined in a similar manner as for PDMPs [64].

Let ω^q be a diffusion trajectory that starts in (q, x^q). Let $t_*(\omega^q)$ be the first hitting time of ∂X^q of the process (x_t^q). Let us define the following *right continuous multiplicative functional*:

$$F\big(t, \omega^q\big) = I_{(t < t_*(\omega^q))} \exp\left[-\int_0^t \lambda\big(q, x_s^q(\omega^q)\big)\, ds\right]. \tag{4.2}$$

This function will be the survivor function for the stopping time S^q associated to the diffusion (x_t^q), which will be employed in the construction of our model. This means that 'killing'[1] of the process (x_t^q) is done according to the multiplicative functional $F(t, \cdot)$ (see [64] for the details about the killed process). The stopping time S^q can be thought of as the minimum of two other stopping times:

1. the first hitting time of boundary, i.e., $t_* |_{\Omega^q}$;
2. the stopping time $S^{q'}$ given by the following continuous multiplicative functional (which plays the role of the survivor function):

$$M\big(t, \omega^q\big) = \exp\left[-\int_0^t \lambda\big(q, x_s^q(\omega^q)\big)\, ds\right].$$

The stopping time $S^{q'}$ can be defined as

$$S^{q'}(\omega^q) = \sup\big\{t \big| \Lambda_t^q(\omega^q) \le m^q(\omega^q)\big\},$$

where Λ_t^q is the following *additive functional* associated to the diffusion (x_t^q):

$$\Lambda_t^q(\omega^q) = \int_0^t \lambda\big(q, x_s^q(\omega^q)\big)\, ds$$

---

[1] In stochastic analysis a killed process is a stochastic process that is forced to assume an undefined or 'killed' state at some (possibly random) time.

and m^q is an $\mathbb{R}_+$-valued random variable on Ω^q, which is exponentially distributed with the survivor function $\mathbb{P}^q_{x^q}[m^q > t] = e^{-t}$. Then

$$\mathbb{P}^q_{x^q}[S^{q'} > t] = \mathbb{P}^q_{x^q}[\Lambda^q_t \le m^q]. \tag{4.3}$$

We set $\omega = \omega^{q_0}$ and the first jump time of the process is

$$T_1(\omega) = T_1(\omega^{q_0}) = S^{q_0}(\omega^{q_0}).$$

The sample path $\mathbf{x}_t(\omega)$ of the hybrid process up to the first jump time is now defined as follows:

(i) If $T_1(\omega) = \infty$ then $\mathbf{x}_t(\omega) = (q_0, x^{q_0}_t(\omega^{q_0}))$, $t \ge 0$.
(ii) If $T_1(\omega) < \infty$ then $\mathbf{x}_t(\omega) = (q_0, x^{q_0}_t(\omega^{q_0}))$, $0 \le t < T_1(\omega)$, where $\mathbf{x}_{T_1}(\omega)$ is a random variable with respect to $R((q_0, x^{q_0}_{T_1}(\omega^{q_0})), \cdot)$.

The process restarts from $\mathbf{x}_{T_1}(\omega) = (q_1, x^{q_1}_1)$ according to the same recipe, using now the process $x^{q_1}_t$. Thus, if $T_1(\omega) < \infty$ we define $\omega = (\omega^{q_0}, \omega^{q_1})$ and the next jump time

$$T_2(\omega) = T_2(\omega^{q_0}, \omega^{q_1}) = T_1(\omega^{q_0}) + S^{q_1}(\omega^{q_1}).$$

The sample path $\mathbf{x}_t(\omega)$ between the two jump times is now defined as follows:

(i) If $T_2(\omega) = \infty$ then $\mathbf{x}_t(\omega) = (q_1, x^{q_1}_{t-T_1}(\omega))$, $t \ge T_1(\omega)$.
(ii) If $T_2(\omega) < \infty$ then $\mathbf{x}_t(\omega) = (q_1, x^{q_1}_t(\omega))$, $0 \le T_1(\omega) \le t < T_2(\omega)$, where $\mathbf{x}_{T_2}(\omega)$ is a random variable with respect to $R((q_1, x^{q_1}_{T_2}(\omega)), \cdot)$.

Then this iteration procedure continues.

We denote

$$N_t(\omega) := \sum I_{(t \ge T_k(\omega))}.$$

$N_t(\omega)$ represents the number of discrete jumps that the trajectory ω records until the time t. The following assumption ensures that the process trajectories do not have chattering (or Zeno executions), i.e., they do not have an infinite number of discrete transitions in a finite amount of time.

Assumption 4.3 (Non-Zeno executions) For every starting point $\mathbf{x} \in \mathbf{X}$,

$$\mathbb{E}N_t < \infty$$

for all $t \in \mathbb{R}_+$.

4.2.3 Formal Definitions

We can define a GSHS as a *hybrid automaton* as follows.

Definition 4.1 A general stochastic hybrid system (GSHS) is a collection

$$H = ((Q, d, \mathscr{X}), b, \sigma, \text{Init}, \lambda, R)$$

where

- Q is a countable/finite set of discrete states;
- $d : Q \to \mathbb{N}$ is a map giving the dimensions of the continuous operation modes;
- $\mathscr{X} : Q \to \mathbb{R}^{d(\cdot)}$ maps each $q \in Q$ into an open subset X^q of $\mathbb{R}^{d(q)}$;
- $b : \mathbf{X}(Q, d, \mathscr{X}) \to \mathbb{R}^{d(\cdot)}$ is a vector field;
- $\sigma : \mathbf{X}(Q, d, \mathscr{X}) \to \mathbb{R}^{d(\cdot) \times m}$ is an $X^{(\cdot)}$-valued matrix, $m \in \mathbb{N}$;
- Init $: \mathscr{B}(\mathbf{X}) \to [0, 1]$ is an initial probability measure on $(\mathbf{X}, \mathscr{B}(\mathbf{X}))$;
- $\lambda : \overline{\mathbf{X}}(Q, d, \mathscr{X}) \to \mathbb{R}^+$ is a transition rate function;
- $R : \overline{\mathbf{X}} \times \mathscr{B}(\mathbf{X}) \to [0, 1]$ is a probabilistic reset kernel.

Following [214], we note that if

- R_c is a transition measure from $(\mathbf{X} \times Q, \mathscr{B}(\mathbf{X} \times Q))$ to $(\mathbf{X}, \mathscr{B}(\mathbf{X}))$; and
- R_d is a transition measure from $(\mathbf{X}, \mathscr{B}(\mathbf{X}))$ to $(Q, \mathscr{B}(Q))$ (where Q is equipped with the discrete topology)

then one might define a probabilistic reset kernel as

$$R(x^q, A) = \sum_{q \in Q} R_d(x^q, q) R_c(x^q, q, A^q)$$

for all $A \in \mathscr{B}(\mathbf{X})$, where $A^q = A \cap (q, X^q)$. Taking this kind of reset map in the definition of a GSHS, the change of the continuous state at a jump depends on the prejump location (continuous and discrete) as well as on the postjump discrete state. This construction can be used to prove that the stochastic hybrid processes with jumps, developed in [26], are a particular class of GSHS executions.

A GSHS execution can be defined as follows.

Definition 4.2 (GSHS execution) A stochastic process $\mathbf{x}_t = (q(t), x(t))$ is called a GSHS execution if there exists a sequence of stopping times $T_0 = 0 < T_1 < T_2 \leq \cdots$ such that for each $k \in \mathbb{N}$,

- $x_0 = (q_0, x_0^{q_0})$ is a $(Q \times X)$-valued random variable extracted according to the probability measure Init;
- for $t \in [T_k, T_{k+1})$, $q_t = q_{T_k}$ is constant and $x(t)$ is a (continuous) solution of the SDE:

$$dx(t) = b(q_{T_k}, x(t)) dt + \sigma(q_{T_k}, x(t)) dW_t \qquad (4.4)$$

  where W_t is the m-dimensional standard Wiener;
- $T_{k+1} = T_k + S^{q_k}$ where S^{q_k} is chosen according to the survivor function (4.2);
- the probability distribution of $x(T_{k+1})$ is governed by the law

$$R((q_{T_k}, x(T_{k+1}^-)), \cdot).$$

We can describe a GSHS giving only its behaviour in terms of random dynamical systems or stochastic hybrid processes. We prefer to present GSHSs as stochastic hybrid automata and to describe their behaviour as a stochastic hybrid process. This presentation corresponds to the usual way of defining hybrid systems in control engineering and computer science.

4.3 Markov String

In this section, we define a very general class of Markov processes, which will be called *Markov strings*, using the mixture operation of Markov processes defined in [182]. A Markov string is a hybrid state 'jump Markov process'. The 'continuous state' component changes its dynamics at random moments among a countable/finite collection of Markov processes defined on some specific evolution modes. The 'discrete component' keeps track of the index of the Markov process that governs the continuous component. This discrete component plays the role of an 'operation index'. The continuous state is allowed to jump whenever the operation index changes. For a Markov string the sojourn time in each operation mode is given as a stopping time with memoryless property of the process that evolves in that mode. Moreover, the continuous state immediately before a switching between modes is allowed to influence that jump.

4.3.1 Informal Description

We start by specifying the main ingredients we need to construct Markov strings:

1. a countable/finite family of independent Markov processes with some good properties (e.g., the strong Markov property, the càdlàg property);
2. a sequence of independent stopping times (a stopping time with memoryless property is given for each process);
3. a probabilistic renewal kernel given a priori.

The given stopping times represent the time moments when the Markov string changes its operation index (i.e., when it jumps from one process to another) and the renewal kernel gives the distribution of the new starting state. The piecing out construction of the Markov string is natural. It is governed by the following algorithm:

Algorithm 4.1   `Initialise the starting state according to the initial probability distribution.`
  `Repeat`

- `From the starting state, follow a trajectory of one process that belongs to the given family;`
- `Stop the above trajectory at the corresponding stopping time;`
- `Reset the current state with a new one according to the renewal kernel;`

  `Until true.`

The pieced-together process obtained by the above procedure is called a Markov string. This section aims to prove that the Markov string inherits good properties (like the strong Markov property and the càdlàg property) from its component processes.

The Markov string construction is closely related to the mixing operation of Markov processes from [182] and the random evolution process construction from [214]. Markov strings differ from the class of processes considered in [182], in that: (i) jump times are essentially given memoryless stopping times, *not necessarily the lifetimes of the component processes*; (ii) after a jump, the string is allowed to restart another process, which might be different from the prejump process.

The mixing ('melange') operation in [182] is only sketched, the author claims that it can be obtained using the renewal ('renaissance') operation. Additionally, Markov strings can be obtained by specialising the base process and the 'instantaneous' distribution in the structure of the random evolution processes developed by Siegrist in [214], but these correspond to GSHSs that allow only spontaneous discrete transitions. There, Siegrist claims that the strong Markov of the composite process can be derived in a manner similar to the strong Markov property of revival processes introduced by Ikeda et al. in [134]. Indeed, for revival processes this property is proved in [134], but a more complete and comprehensible proof is given by Meyer in [182]. For Markov strings, a very elaborate proof for the strong Markov property was given in [28]. In this chapter, we underline a rather simpler proof for this property. Intuitively, because the component processes are strong Markov and the piecing out mechanism is based on memoryless stopping times, the whole Markov string should have again the strong Markov property. In fact, we will suppose some supplementary hypotheses about the component processes, like the fact they are right Markov processes (see the following section) and their trajectories are càdlàgs, so the description of the piecing out mechanism may be much more easily understood. The resulting Markov string will also be a right process, and then, the strong Markov property can be derived just as a simple consequence. These extra hypotheses are motivated by the fact that for a GSHS, continuous dynamics in the modes is not very complicated. They may be described by SDE or SDE with jumps. Therefore, for the scope of this book, we do not need to investigate properties for stochastic processes that result from pasting together very general Markov processes.

4.3.2 The Components

Suppose that we have an indexed countable/finite family of Markov processes

$$M^q = \left(\Omega^q, \mathscr{F}^q, \mathscr{F}_t^q, x_t^q, \theta_t^q, \mathbb{P}^q, \mathbb{P}_{x^q}^q \right), \quad q \in Q.$$

For each M^q, the state space is described by $(X^q, \mathscr{B}^q)$, where $\mathscr{B}^q$ is the Borel σ-algebra of X^q if the latter is a topological Hausdorff space. Suppose that Δ is the cemetery point for all X^q, $q \in Q$. For each $q \in Q$, the elements $\mathscr{F}^q$, $\mathscr{F}_t^{q,0}$, $\mathscr{F}_t^q$, θ_t^q, $\mathbb{P}^q$, $\mathbb{P}_{x^q}^q$ have the usual meaning in Markov process theory. $(\mathbf{P}_t^q)$ denotes the operator semigroup associated to M^q, which maps $\mathscr{B}_b^q(X^q)$ into itself, given by

$$\mathbf{P}_t^q f^q(x^q) = \mathbb{E}_{x^q}^q f^q(x_t^q),$$

where $\mathbb{E}_{x^q}^q$ is the expectation with respect to $\mathbb{P}_{x^q}^q$. For M^q, a function $f^q \geq 0$ is p-excessive ($p > 0$) if $e^{-pt}\mathbf{P}_t^q f^q \leq f^q$ for all $t \geq 0$ and $e^{-pt}\mathbf{P}_t^q f^q \nearrow f^q$ as $t \searrow 0$. For each, $q \in Q$, we may also consider the operator resolvent (V_α^q) associated to M^q and define the excessive functions with respect to this resolvent.

The main assumptions regarding the processes $\{M^q, q \in Q\}$ state that these processes satisfy the *hypothèses droites* denoted HD1 and HD2, defined by Meyer in [181], i.e., they are *right processes*. These hypotheses have subsequently undergone various subtle modifications, and the reader may find a complete study in [208]. In principle, a right process is a strong Markov process whose trajectories are almost sure right continuous and the operator semigroup (or resolvent) satisfies some technical conditions.

In this section, we impose some extra assumptions on the processes M^q, so they may be easily specified and handled. These are motivated by stochastic hybrid system models that do not require very general topological state spaces or require only right continuity of the paths. In most cases, the state spaces can be embedded in some Euclidean spaces, and the trajectories are continuous or right continuous with left limits.

In the following assumption, we summarise all the hypotheses regarding component processes.

Assumption 4.4 For each $q \in Q$, we suppose the following.

(i) The p-excessive functions of M^q are $\mathbb{P}^q$-a.s. right continuous on trajectories.
(ii) $\mathbb{P}^q$ is a complete probability.
(iii) The state space X^q is a Borel space.
(iv) Trajectories of M^q have the càdlàg property, i.e., for each $\omega^q \in \Omega^q$, the sample path $t \mapsto x_t^q(\omega^q)$ is right continuous on $[0, \infty)$ and has left limits on $(0, \infty)$ (inside X_Δ^q).

Part (iv) of the assumption implies that the underlying probability space Ω^q can be assumed to be $D_{[0,\infty)}(X^q)$, the space of functions mapping $[0, \infty)$ to X^q, which are right continuous functions with left limits (càdlàgs). Let us consider ω_Δ^q the cemetery point of Ω^q corresponding to the 'dead' trajectory of M^q (when the process is trapped to Δ).

The condition (i) implies that the strong Markov property is also satisfied:

(i') M^q is a strong Markov process.

According to [181], Assumption 4.4 implies that each M^q is also a *right process*. We can assume, without loss of generality, that

$$X^q \cap X^{q'} = \emptyset$$

if $q \neq q'$. Then we derive that

$$\Omega^q \cap \Omega^{q'} = \emptyset$$

for $q \neq q'$.

At this point, we have to emphasise the fact that, using the family $\{M^q\}_{q \in Q}$, we can construct a rather general process, which is the superposition of the processes M^q,

$$\widehat{M} := \bigoplus_{q \in Q} M^q.$$

This process simultaneously keeps track of each process dynamics together with the rule that properly 'generates' this dynamics (see Fig. 4.1). Then

$$\widehat{M} := (\widehat{\mathbf{x}}_t) = (q, x_t^q).$$

The state space $\mathbf{X}$ is defined as follows. $\mathbf{X}$ is constructed as the direct sum of spaces X^q, with the same deadlock point Δ, i.e.,

$$\mathbf{X} = \bigcup_{q \in Q} \{(q, x) | x \in X^q\}. \tag{4.5}$$

In the same manner as in Sect. 4.2, $\mathbf{X}$ is a Borel space.

$\mathbf{X}$ can be endowed with the Borel σ-algebra $\mathscr{B}(\mathbf{X})$ generated by the topology that is generated by its metric. Moreover, we have

$$\mathscr{B}(\mathbf{X}) = \sigma \left\{ \bigcup_{q \in Q} \{q\} \times \mathscr{B}^q \right\}. \tag{4.6}$$

Then $(\mathbf{X}, \mathscr{B}(\mathbf{X}))$ is a Borel space, whose Borel σ-algebra $\mathscr{B}(\mathbf{X})$ restricted to each component X^q gives the initial σ-algebra $\mathscr{B}^q$ [64].

Since $X^q \cap X^{q'} = \emptyset$ if $q \neq q'$, the relations (4.5) and (4.6) become

$$\mathbf{X} = \bigcup_{q \in Q} X^q, \tag{4.7}$$

$$\mathscr{B}(\mathbf{X}) = \sigma \left(\bigcup_{q \in Q} \mathscr{B}^q \right). \tag{4.8}$$

We use, alternatively, the notation

$$\mathbf{x} := (q, x) = x^q.$$

We denote

$$\widehat{\Omega} := \bigcup_{q \in Q} \Omega^q.$$

An arbitrary element of Ω (respectively Ω^q) is denoted by $\widehat{\omega}$ (respectively ω^q). Then, if $\widehat{\omega} = \omega^q \in \Omega^q$, we have

$$\widehat{\mathbf{x}}_t(\widehat{\omega}) := (q, x_t^q(\omega^q))$$

Fig. 4.1 Process
superposition

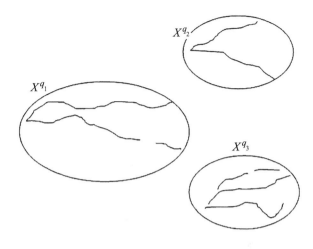

and

$$\widehat{\theta}_t(\widehat{\omega}) := \theta_t^q(\omega^q).$$

The filtration for $\widehat{M}$ will be defined as follows:

$$\widehat{\mathscr{F}} := \big\{ \Gamma \subseteq \widehat{\Omega} \big| \Gamma \cap \Omega^q \in \mathscr{F} \text{ for } q \in Q \big\},$$

$$\widehat{\mathscr{F}}_t := \big\{ \Gamma \subseteq \widehat{\Omega} \big| \Gamma \cap \Omega^q \in \mathscr{F}_t \text{ for } q \in Q \big\}.$$

The σ-algebra $\widehat{\mathscr{F}}$ can be defined also as $\sigma(\bigcup_{q \in Q} \mathscr{F}^q)$ on $\widehat{\Omega}$.
Finally, the Wiener probabilities are

$$\widehat{\mathbb{P}}_{\mathbf{x}}(\Gamma) := \mathbb{P}_{x^q}^q(\Gamma \cap \Omega^q).$$

Remark 4.3 Based, on the previous definitions, it is easy to check that:

(i) $(\widehat{x}_t, \widehat{\mathbb{P}}_{x^q})$ is equivalent to $((q, x_t^q), \mathbb{P}_{x^q}^q)$ for $x^q \in \mathbf{X}$;
(ii) $\widehat{M}$ inherits the properties of the constituent Markov processes like strong
   Markov, right continuous, left limits or other continuities on the paths.

In the remainder of this section, our purpose is to construct a new stochastic
process that is subordinated to $\widehat{M}$ and can be also obtained by sequential composition of the realisations of some specific subprocesses $\widetilde{M}^q$ of M^q. The subprocesses
$\widetilde{M}^q$ result from the processes M^q by killing (stopping) them after some given stopping times. With this in mind, we need to define the transition mechanism between
two consecutive processes. This transition mechanism must give details about the
switching/jumping time and the new location where the new process is starting up.
Formally, we need:

1. a stopping time (which gives the jump temporal parameter) for each process;
2. a renewal probabilistic kernel, which gives the postjump state.

Therefore, to define the required Markov string M, the following mathematical
objects are needed:

1. for each $q \in Q$, S^q is a *stopping time* of M^q;
2. a *renewal kernel* that is a Markovian kernel

$$\Psi : \left\{ \bigcup_{q \in Q} \Omega^q \right\} \times \mathscr{B}(\mathbf{X}) \to [0, 1].$$

The following assumption states the necessary properties of these objects.

Assumption 4.5

(i) For each $q \in Q$, S^q is terminal time, i.e., stopping time with the 'memoryless' property:

$$S^q \left(\theta_t^q \omega^q \right) = S^q \left(\omega^q \right) - t, \quad \forall t < S^q \left(\omega^q \right). \tag{4.9}$$

(ii) The renewal kernel Ψ satisfies the following conditions:
  (a) if $S^q (\omega^q) = +\infty$ then $\Psi (\omega^q, \cdot) = \varepsilon_\Delta$ (here, ε_Δ is the Dirac measure corresponding to Δ);
  (b) if $t < S^q (\omega^q)$ then

$$\Psi \left(\theta_t^q \omega^q, \cdot \right) = \Psi \left(\omega^q, \cdot \right). \tag{4.10}$$

Note that the component processes have the càdlàg property, therefore they may also have jumps, which are not treated separately in the construction of the Markov strings. The sequence of jump times refers to additional jumps, not to the discontinuities of the trajectories of component processes.

We consider now, for each $q \in Q$, the killed process

$$\tilde{M}^q = \left(\Omega^q, \mathscr{F}^q, \mathscr{F}_t^q, \tilde{x}_t^q, \tilde{\theta}_t^q, \mathbb{P}^q, \mathbb{P}_{x^q}^q \right)$$

after S^q, where

$$\tilde{x}_t^q \left(\omega^q \right) = \begin{cases} x_t^q (\omega^q), & \text{if } t < S^q (\omega^q) \\ \Delta, & \text{if } t \geq S^q (\omega^q) \end{cases}$$

and

$$\tilde{\theta}_t^q \left(\omega^q \right) = \begin{cases} \theta_t^q (\omega^q), & \text{if } t < S^q (\omega^q) \\ \omega_\Delta^q, & \text{if } t \geq S^q (\omega^q). \end{cases}$$

In this case, Ω^q should be thought of as a subspace of $\Omega^q \times [0, \infty)$; the above embedding is made through the map

$$\omega^q \mapsto \left(\omega^q, S^q \left(\omega^q \right) \right).$$

The killed process is equivalent with the subprocess of M^q corresponding to the multiplicative functional

$$\alpha^q \left(t, \omega^q \right) = I_{[0, S^q (\omega^q))}(t) \tag{4.11}$$

(see Chap. III [33]).

Fig. 4.2 Subprocess
superposition

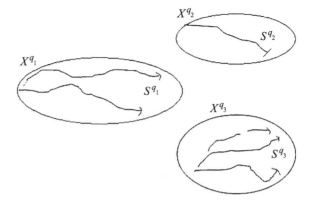

Using the family (α^q) of multiplicative functionals, we can also define the following multiplicative functional for the superposition process $\widehat{M}$:

$$\widehat{\alpha}(t, \widehat{\omega}) = \alpha^q \left(t, \omega^q\right) = I_{[0, S^q(\omega^q))}(t)$$

for $\widehat{\omega} = \omega^q \in \Omega^q$.

The canonical subprocess (see Fig. 4.2)

$$\widetilde{M} = (\widetilde{\Omega}, \widetilde{\mathscr{F}}, \widetilde{\mathscr{F}}_t, \widetilde{\mathbf{x}}_t, \widetilde{\theta}_t, \widetilde{\mathbb{P}}_{\mathbf{x}})$$

is a right Markov process associated with $\widehat{M}$ and $\widehat{\alpha}$ with the state space $(\mathbf{X}, \mathscr{B})$ (augmented with Δ) such that

$$\mathbb{E}_{\mathbf{x}}\left[f(\widetilde{\mathbf{x}}_t)\right] = \mathbb{E}_{x^q}^q\left[f^q\left(x_t^q\right)I_{[0, S^q)}(t)\right]$$

for all bounded measurable functions

$$f = \bigoplus f^q$$

and

$$\mathbf{x} = x^q \in \mathbf{X}$$

(see Chap. II [208]). The sample paths of $\widetilde{M}$ coincide with the sample paths of $\widehat{M}$ up to the random time

$$\widetilde{\zeta} := \bigoplus S^q,$$

after which $\widetilde{M}$ is trapped in the cemetery state Δ. Therefore, $\widetilde{M}$ can be thought of as the superposition of the subprocesses $\widetilde{M}^q$.

4.3.3 Piecing out Markov Processes

In this section, the elements defined in the Sect. 4.3.2 will be further employed to construct the sequentially pieced-together stochastic process

Fig. 4.3 Markov string

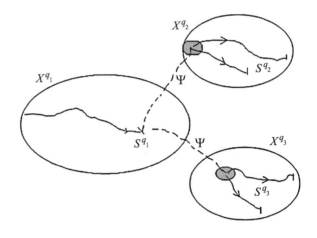

$$M = (\Omega, \mathscr{F}, \mathscr{F}_t, \mathbf{x}_t, \theta_t, \mathbb{P}, \mathbb{P}_x),$$

which will be called a *Markov string*. This new process M is obtained by the concatenation of the paths of the subprocesses $\widetilde{M}^q$ (see Fig. 4.3).

For a complete definition of the Markov string, as a stochastic process, one needs to specify all its parameters:

1. $(\mathbf{X}, \mathscr{B})$: the state space;
2. $(\Omega, \mathscr{F}, \mathscr{F}_t, \mathbb{P}, \mathbb{P}_x)$: the underlying probability filtered space;
3. θ_t: the translation/shift operator.

Without loss of generality, we take for M the same state space $(\mathbf{X}, \mathscr{B})$ as we did for the superposition process $\widehat{M}$.

The event space Ω can be thought of as the space generated by the concatenation operation defined on the union of the spaces Ω^q (which are pairwise disjoint), i.e.,

$$\Omega := \left(\bigcup_{q \in Q} \Omega^q \right)^*.$$

Note that, for each $q \in Q$, an arbitrary element ω^q of Ω^q must be thought of as a trajectory of the subordinated process $\widetilde{M}^q$. The cemetery point of Ω is denoted by

$$\omega_\Delta = \left(\omega_\Delta^q \right)_{q \in Q}.$$

An arbitrary element of Ω is denoted by ω.

The σ-algebra $\mathscr{F}$ on Ω will be the smallest σ-algebra on Ω such that the projections

$$\pi^q : \Omega \to \Omega^q$$

are $\mathscr{F}/\mathscr{F}^q$-measurable, $q \in Q$. The probability $\mathbb{P}$ on $\mathscr{F}$ will be defined as a 'product measure'.

An alternative construction for Ω can be done as in [134]. There, the core ideas consider taking the elements of the state space $\mathbf{X}$ to 'label' the events of Ω. First, we take

$$\overline{\Omega} := \widetilde{\Omega} \times \mathbf{X}_\Delta$$

with the σ-algebra

$$\overline{\mathscr{F}} := \widetilde{\mathscr{F}} \otimes \mathscr{B}(\mathbf{X}_\Delta).$$

In the second step, we define the probability space as

$$\Omega := \prod_{j=1}^{\infty} \overline{\Omega}_j,$$

where $\overline{\Omega}_j = \overline{\Omega}$, $j = 1, 2, \ldots$, with the product σ-algebra

$$\mathscr{F} := \bigotimes_{j=1}^{\infty} \overline{\mathscr{F}}_j,$$

where $\overline{\mathscr{F}}_j = \overline{\mathscr{F}}$, $j = 1, 2, \ldots$. On $(\overline{\Omega}, \overline{\mathscr{F}})$, using the renewal kernel Ψ, we can set up the following set of Wiener probabilities:

$$\mathbb{Q}_x(d\overline{\omega}) := \widetilde{\mathbb{P}}_x(d\widetilde{\omega})\Psi(\widetilde{\omega}, \mathbf{x}),$$

where we denote

$$\overline{\omega} := (\widetilde{\omega}, \mathbf{x}).$$

In this way, we can use Ionescu–Tulcea's theorem to define the Wiener probabilities $\mathbb{P}_\mathbf{x}$ on $(\Omega, \mathscr{F})$ that must satisfy the following condition:

$$\mathbb{P}_\mathbf{x}[d\overline{\omega}_1, d\overline{\omega}_2, \ldots, d\overline{\omega}_n] = \mathbb{Q}_{\mathbf{x}_1}(d\overline{\omega}_1)\mathbb{Q}_{\mathbf{x}_2}(d\overline{\omega}_2)\ldots\mathbb{Q}_{\mathbf{x}_n}(d\overline{\omega}_n)$$

where $\overline{\omega}_i = (\widetilde{\omega}, \mathbf{x}_i)$.

The natural filtration $\mathscr{F}_t$ will be associated in a standard way to the Markov string.

Piecing out Algorithm   Now we give a step-by-step methodology to construct a sample path of the stochastic process $(\mathbf{x}_t)_{t>0}$ with values in $\mathbf{X}$, starting from a given initial state

$$x_0 = x_0^{q_0} \in X^{q_0}.$$

Practically, our method to construct a Markov string starting is nothing but the description of a Markov chain with values in the space of paths $\widehat{\Omega}$ starting with the initial path ω^{q_0}.

Let ω^{q_0} be a sample path of the process $(x_t^{q_0})$ starting with x_0.

Let

$$T_1(\omega^{q_0}) = S^{q_0}(\omega^{q_0}).$$

The event ω and the associated sample path are inductively defined. Initially, we set

$$\omega := \omega^{q_0}.$$

Then the sample path $\mathbf{x}_t(\omega)$ up to the first jump time coincides with ω^{q_0}, i.e., is defined as follows:

(i) if $T_1(\omega) = \infty$ then $\mathbf{x}_t(\omega) = x_t^{q_0}(\omega^{q_0})$, $t \geq 0$;
(ii) if $T_1(\omega) < \infty$ then $\mathbf{x}_t(\omega) = x_t^{q_0}(\omega^{q_0})$, $0 \leq t < T_1(\omega)$, where $\mathbf{x}_{T_1}$ is a random variable chosen according to $\Psi(\omega^{q_0}, \cdot)$.

The process $(\mathbf{x}_t)$ restarts from $\mathbf{x}_{T_1} = x_1^{q_1}$ following now the dynamics rules of the process $M^{q_1} = (x_t^{q_1})$. Let ω^{q_1} be a sample path of the process M^{q_1} starting with $x_1^{q_1}$. Thus, provided that $T_1(\omega) < \infty$, we define the next jump time

$$T_2(\omega^{q_0}, \omega^{q_1}) := T_1(\omega^{q_0}) + S_{q_2}(\omega^{q_1}).$$

Then in the second step we 'record' the swept path

$$\omega := \omega^{q_0} * \omega^{q_1},$$

where $*$ is the concatenation operation of trajectories. Between the two jump times, $\mathbf{x}_t(\omega)$ coincides with $(x_t^{q_1})$:

(i) if $T_2(\omega) = \infty$ then $\mathbf{x}_t(\omega) = x_{t-T_1}^{q_1}(\omega^{q_1})$, $t \geq T_1(\omega)$;
(ii) if $T_2(\omega) < \infty$ then $\mathbf{x}_t(\omega) = x_t^{q_1}(\omega^{q_1})$, $0 \leq T_1(\omega) \leq t < T_2(\omega)$, where $\mathbf{x}_{T_2}$ is a random variable chosen according to $\Psi(\omega^{q_1}, \cdot)$.

Generally, if $T_k(\omega) = T_k(\omega^{q_0}, \omega^{q_1}, \ldots, \omega^{q_{k-1}}) < \infty$ with

$$\omega := \omega^{q_0} * \omega^{q_1} * \cdots * \omega^{q_{k-1}}$$

then the next jump time is

$$T_{k+1}(\omega) = T_{k+1}(\omega^{q_0}, \omega^{q_1}, \ldots, \omega^{q_k}) = T_k(\omega^{q_0}, \omega^{q_1}, \ldots, \omega^{q_{k-1}}) + S^{q_k}(\omega^{q_k}).$$
$$(4.12)$$

Between the two jump times T_k and T_{k+1}, the trajectory $\mathbf{x}_t(\omega)$ evolves according to the rule of the corresponding Markov component, as follows:

(i) if $T_{k+1}(\omega) = \infty$ then $\mathbf{x}_t(\omega) = x_{t-T_k}^{q_k}(\omega^{q_k})$, $t \geq T_{k+1}(\omega)$;
(ii) if $T_{k+1}(\omega) < \infty$ then $\mathbf{x}_t(\omega) = x_{t-T_k}^{q_k}(\omega^{q_k})$, $0 \leq T_k(\omega) \leq t < T_{k+1}(\omega)$, where $\mathbf{x}_{T_{k+1}}$ is a random variable chosen according to $\Psi(\omega^{q_k}, \cdot)$.

We have constructed a sequence of jump times

$$0 < T_1 < T_2 < \cdots < T_n < \cdots.$$

Let $T_\infty = \lim_{n \to \infty} T_n$. Then

$$\mathbf{x}_t(\omega) = \Delta \quad \text{if } t \geq T_\infty.$$

Define

$$k_0 := \min\{k : S^{q_k}(\omega) = \infty\}. \tag{4.13}$$

The process $(\mathbf{x}_t)$ starting with $\mathbf{x}_0 = (q_0, x_0^{q_0})$ until T_{k_0} is a 'particle' that moves on the following composed trajectory:

$$\omega = \omega^{q_0} * \omega^{q_1} * \cdots * \omega^{q_{k_0}-1}.$$

We denote

$$N_t(\omega) = \sum I_{(t \geq T_k)}$$

the number of jump times in the interval $[0, t]$. To eliminate pathological solutions that take an infinite number of discrete transitions in a finite amount of time (known as *Zeno solutions*) we impose the following assumption:

Assumption 4.6 (Non-Zeno dynamics) For every starting point $\mathbf{x} \in \mathbf{X}$,

$$\mathbb{E}_{\mathbf{x}} N_t < \infty$$

for all $t \in \mathbb{R}_+$.

Under Assumption 4.6, the underlying probability space Ω can be identified with $D_{[0,\infty)}(\mathbf{X})$.

Shift Operator   The shift operator θ_t of $\omega \in \Omega$ is defined as

$$\theta_t \omega := \left(\theta_{t-T_k}^{q_k} \left(\omega^{q_k} \right), \omega^{q_{k+1}}, \ldots \right) \quad \text{if } T_k(\omega) \leq t < T_{k+1}(\omega),$$

where $\omega = (\omega^{q_0}, \omega^{q_1}, \ldots)$.

An alternative way to formally present these mathematical objects is given in [28]. We may observe that these definitions belonging to the Markov process arsenal are quite technical and lengthy, but the intuition behind them is clear. The reader who is not interested in a deep study may skip them, keeping in mind that, locally, the dynamics of a Markov string is, in fact, the dynamics of one of its component processes.

4.3.4 Basic Properties

The main objective of this section is to prove that the Markov string $(\mathbf{x}_t)$ constructed in Sect. 4.3.3 is a *right Markov process*. The proof skeleton is given by the Markov property of the discrete time Markov chain (p_n), which is constructed on the union of sets of trajectories of the component Markov processes and uses the renewal kernel to define its transition probability.

Concretely, (p_n) is a discrete time Markov chain with the state space $(\widehat{\Omega}, \widehat{\mathscr{F}})$ and the underlying probability space $(\Omega, \mathscr{F})$ that describes the jump structure of $(\mathbf{x}_t)$. The chain state p_n is essentially the 'nth' step of the process $(\mathbf{x}_t)$. If p_n starts in ω^{q_0} (an element in Ω^{q_0} commencing in $x_0^{q_0}$) then $p_n(\omega) = \omega^{q_n}$.

The transition kernel associated with (p_n) can be defined as follows:

$$H(\widehat{\omega}, A) = \mathbb{P}_\psi(\widehat{\omega}, A), \quad A \in \widehat{\mathscr{F}}.$$

The construction of $\mathbb{P}_{\mathbf{x}}$ of the Markov string is such that

- H is the transition function of (p_n);
- $\mathbb{P}_{\widehat{\omega}}$ is the initial probability law of (p_n); i.e., if $\widehat{\omega} \in \widehat{\Omega}$, which starts in $\mathbf{x} \in \mathbf{X}$, then

$$\mathbb{P}_{\widehat{\omega}}(p_0 \in A) = \mathbb{P}_{\mathbf{x}}(A), \quad A \in \mathscr{F}.$$

Let η_k be the projection $(p_0, p_1, \ldots, p_k)$, i.e.,

$$\eta_k := (p_0, p_1, \ldots, p_k),$$
$$\eta_k(\omega) := \left(\omega^{q_0} * \omega^{q_1} * \cdots * \omega^{q_k}\right).$$

The Markov chain (p_n) can be further employed in the study of the Markov string $(\mathbf{x}_t)$. For each k on the set $\{T_k(\omega) \le t < T_{k+1}(\omega)\}$ we have

$$\mathbf{x}_t = x_{t-T_k}^{q_k} \circ p_k.$$

We will call Assumptions 4.4, 4.5 and 4.6 *standard assumptions for a Markov string*. The next result states the most important properties of Markov strings.

Theorem 4.4 *Under the standard assumptions, any Markov string*

$$M = (\Omega, \mathscr{F}, \mathscr{F}_t, \mathbf{x}_t, \theta_t, \mathbb{P}, \mathbb{P}_{\mathbf{x}})$$

has the following properties:

  (i) *is a strong Markov process*;
 (ii) *has the càdlàg property*;
(iii) *is a right process*.

Clearly, condition (iii) implies condition (i) in the above theorem. The statement of this theorem will be proved as a consequence of some auxiliary results described below. In principle, Theorem 4.4 shows that Markov strings inherit the properties of their components.

Proposition 4.5 (Càdlàg property) *Under the standard assumptions, any Markov string has the càdlàg property, i.e., for all $\omega \in \Omega$ the trajectories $t \mapsto \mathbf{x}_t(\omega)$ are right continuous on $[0, \infty)$ with left limits on $(0, \infty)$.*

Proof The result is a direct consequence of two facts:

1. The sample paths of $(\mathbf{x}_t)$ are obtained by the concatenation of sample paths of component process (i.e., the concatenation is done in a way that preserves the right continuity and the left limits).
2. The component processes are taken such that their trajectories have the càdlàg property.

Then the Markov string inherits the càdlàg property. $\square$

Proposition 4.6 *Under standard assumptions, any Markov string is a right process.*

Proof The proof can be done by induction, considering the fact that the number of components for a Markov string is at most countable. The following lemma is essential in proving the induction steps.   □

Lemma 4.7 *Any Markov string with two components M^1 and M^2 (i.e., card$(Q) = 2$) is a right Markov process.*

Proof For each M^q, the state space is given by $(X^q, \mathcal{B}^q)$, $q = 1, 2$. Consider the jumping times from one process to another as the sequence

$$T_0 = 0 < T_1 < T_2 < \cdots < T_n < \cdots$$

constructed as in the piecing out algorithm using the terminal times S^1 and S^2.

Let $\mathbf{x} = (q, x)$ be an arbitrary state in $\mathbf{X}$ and f a measurable bounded function on $\mathbf{X}$. We make the following notations:

$$f^q := f \cdot 1_{X^q} = f|_{X^q}, \quad q = 1, 2,$$
$$T_{i,i+1}(\omega) := \{t \,|\, T_i(\omega) \le t < T_{i+1}(\omega)\}, \quad i = 0, 1, 2, \ldots.$$

In the interval time $T_{i,i+1}(\omega)$, the evolution mode index q where the string evolves is denoted by $\mathrm{ind}(i)$, i.e.,

$$\mathrm{index}\big[T_{i,i+1}(\omega)\big] := \mathrm{ind}(i).$$

Then

$$T_{i+1}(\omega) - T_i(\omega) = S^{\mathrm{ind}(i)}.$$

The transition semigroup (expectations with respect to the Wiener probabilities) for the composed process is

$$\mathbf{P}_t f(\mathbf{x}) := \mathbb{E}_{\mathbf{x}} f(\mathbf{x}_t)$$
$$= \sum_{q=1}^{2} \left\{ 1_{X^q}(x) \cdot \mathbb{E}_x^q \sum_{i=0}^{\infty} \big[1_{T_{i,i+1}}(t) \cdot \Psi^i \big[\mathbf{P}_{t-T_i}^{\mathrm{ind}(i)} \big(f^{\mathrm{ind}(i)} \big) \big] \big] \right\}, \quad (4.14)$$

where

$$\Psi^i = \underbrace{\Psi \circ \cdots \circ \Psi}_{i \text{ times}}$$

is the i-times iterated action of kernel Ψ on the measurable bounded function $\mathbf{P}_{t-T_i}^{\mathrm{ind}(i)}(f^{\mathrm{ind}(i)})$. Recall that the action of a stochastic kernel on a measurable function is the integral of that function with respect to the probabilistic measure defined by the kernel in the second argument.

Intuitively, the semigroup formula for a function f (4.14) says that if the string starts in a particular state (q, x), then $\mathbf{P}_t f(\mathbf{x})$ is the sum of all terms that result from applying expectation $\mathbb{E}_x^q$ to all measurable 'piece' functions defined on X^q obtained from f through the discrete and continuous transitions of the entire string.

Now, even though we have not yet verified that $(\mathbf{P}_t)$ is a semigroup, we set its Laplace transform (resolvent):

$$V_\alpha := \int_0^\infty e^{-\alpha t} \mathbf{P}_t \, dt.$$

Since $(\mathbf{x}_t)$ has the càdlàg property, it is, in particular, right continuous; then it suffices to prove (according to [208]) that, for fixed $t \geq 0$, $\mathbf{x} \in \mathbf{X}$, and $f \in C(\mathbf{X})$ ($C(\mathbf{X})$ is the space of continuous real functions on $\mathbf{X}$), the following two assertions hold:

$$\mathbb{E}_\mathbf{x}\left\{ \int_0^\infty e^{-\alpha r} f(\mathbf{x}_{t+r}) \, dr \cdot J \right\} = \mathbb{E}_\mathbf{x}\{ V_\alpha f(\mathbf{x}_t) \cdot J \}, \tag{4.15}$$

$$t \mapsto V_\alpha f(\mathbf{x}_t) \quad \text{is a.s. right continuous} \tag{4.16}$$

for all J that are bounded and $\mathscr{F}_t$-measurable. Practically, (4.15) and (4.16) mean that the process is simple Markov, and, respectively, strong Markov. The idea is that (4.15) is the Laplace transform of the simple Markov property, and this relation can be inverted considering the right continuity of the semigroup. Assertion (4.16) is equivalent to the strong Markov property [208].

For $\mathbf{x} = (q, x)$, let us take the Laplace transform of (4.14) as follows:

$$V_\alpha f(\mathbf{x}) = \sum_{q=1}^2 \left\{ 1_{X^q}(x) \cdot \mathbb{E}_x^q \sum_{i=0}^\infty \left[\Psi^i(\cdot, dy) \int_{T_i}^{T_{i+1}} e^{-\alpha t} \cdot \mathbf{P}_{t-T_i}^{\text{ind}(i)} \left(f^{\text{ind}(i)} \right)(y) \, dt \right] \right\}$$

$$= \sum_{q=1}^2 \left\{ 1_{X^q}(x) \cdot \mathbb{E}_x^q \sum_{i=0}^\infty \left[e^{-\alpha T_i} \Psi^i(\cdot, dy) \right.\right.$$

$$\left.\left. \cdot \int_0^{S^{\text{ind}(i)}} e^{-\alpha t} \cdot \mathbf{P}_t^{\text{ind}(i)} \left(f^{\text{ind}(i)} \right)(y) \, dt \right] \right\}$$

$$= \sum_{q=1}^2 \left\{ 1_{X^q}(x) \cdot \sum_{i=0}^\infty \mathbb{E}_x^q \left[e^{-\alpha T_i} \Psi^i(\cdot, dy) \widetilde{V}_\alpha^{\text{ind}(i)} \left(f^{\text{ind}(i)} \right)(y) \right] \right\}. \tag{4.17}$$

Here, we denote by $\widetilde{V}_\alpha^q$ the resolvent of the subprocess $\widetilde{M}^q$ (that is, the process M^q killed after the terminal time S^q). Let

$$g^i(\mathbf{x}) := \mathbb{E}_x^q \left[e^{-\alpha T_i} \Psi^i \, \widetilde{V}_\alpha^{\text{ind}(i)} \left(f^{\text{ind}(i)} \right) \right].$$

By using the shift property (4.10) of the renewal kernel Ψ and the memoryless property of the jumping times (4.9), we obtain

$$e^{-\alpha t} \mathbf{P}_t^{\text{ind}(i)} g^i(\mathbf{x}) = \mathbb{E}_x^q \left[e^{-\alpha t} e^{-\alpha (T_i \circ \theta_t^{\text{ind}(i)})} \left(\Psi^i \, \widetilde{V}_\alpha^{\text{ind}(i)} \left(f^{\text{ind}(i)} \right) \right) \circ \theta_t^{\text{ind}(i)} \cdot 1_{[t < S^{\text{ind}(i)}]} \right]$$

$$= \mathbb{E}_x^q \left[e^{-\alpha T_i} \Psi^i \, \widetilde{V}_\alpha^{\text{ind}(i)} \left(f^{\text{ind}(i)} \right) \cdot 1_{[t < S^{\text{ind}(i)}]} \right].$$

From this expression, it is clear that g^i is α-excessive for $\widetilde{M}^{\mathrm{ind}(i)}$. Therefore, since

$$V_\alpha f(\mathbf{x}) = \sum_{q=1}^{2} 1_{X^q}(x) \cdot \sum_{i=0}^{\infty} g^i(\mathbf{x}) \tag{4.18}$$

and the sum in the right hand side of (4.18) is finite (the string has to stop jumping from one component to another), $V_\alpha f$ is a.s. right continuous, i.e., the assertion (4.16) is proved.

In order to prove the simple Markov property, we can prove (4.15) for each random interval $T_{i,i+1}$. Observe that

$$\int_0^\infty e^{-\alpha r} f\left(\mathbf{x}_{t+r}(\omega)\right) dr$$

$$= \sum_{i=0}^{\infty} 1_{T_{i,i+1}}(t) \int_0^\infty e^{-\alpha r} f^{\mathrm{ind}(i)}\left(x_{t+r-T_i(\omega)}^{\mathrm{ind}(i)}\left(\omega^{\mathrm{ind}(i)}\right)\right) dr$$

$$= \sum_{i=0}^{\infty} 1_{T_{i,i+1}}(t) \left(\int_0^\infty e^{-\alpha r} f^{\mathrm{ind}(i)}\left(x_r^{\mathrm{ind}(i)}\right) dr \right) \left(\theta_{t-T_i}^{\mathrm{ind}(i)}\left(\omega^{\mathrm{ind}(i)}\right)\right). \tag{4.19}$$

Note that in the previous equalities, T_i does not depend on $\omega^{\mathrm{ind}(i)}$.

Then, because of the definition of $\mathbb{P}_{\mathbf{x}}$, it is enough to treat the case when the function J depends only on a finite number of variables, i.e.,

$$J(\omega) = J\left(\omega^{q_0} * \omega^{q_1} * \cdots * \omega^{q_k}\right), \quad k \geq 0. \tag{4.20}$$

Now, the simple Markov property can be obtained plugging (4.20), and (4.19) into the left hand side of (4.15) and then employing the Markov property of the process $(x_r^{\mathrm{ind}(i)})$ at the instance $t - T_i$, and the expression of the string resolvent (4.17). □

Remark 4.8 A more direct proof for the simple Markov property of the Markov strings is given in [28]. However, the intuitions behind these proofs are clear: due to the memoryless of the mechanism to go back and forth between process components, the Markov property is inherited by the whole Markov string. But the mathematical formalism is quite heavy making use of so many labels, indices corresponding to the different string paths, and so on; and it shades down these intuitions. In the following section, we will point out the renewal equation that characterises the transition semigroup of a Markov string. This is more transparent and describes the Markovianity inheritance.

4.4 Stochastic Hybrid Processes

Suppose that the GSHS

$$H = \left((Q, d, \mathcal{X}), b, \sigma, \mathrm{Init}, \lambda, R\right)$$

satisfies the standard Assumptions 4.1, 4.2 and 4.3. Then its realisation can be thought of as a particular Markov string, where the components are diffusion processes. Considering that the continuous evolution of such string changes according to different stochastic rules in different modes, this Markov string is often called a *stochastic hybrid process*. We prefer this name instead of any other possible names like 'diffusions with regime switching', or 'piecewise diffusion processes' or 'hybrid switching diffusion processes' that are sometimes met in stochastic calculus/control literature. The motivation for this is the fact that in the Markov string structure, one may use combination of different types of process: diffusions, jump-diffusions, dynamical systems, Markov jump processes, etc. Therefore, we have the freedom not necessarily to use the same type of process. The result will be nothing else but a 'general' stochastic hybrid process.

Let us denote by M_H the realisation of a GSHS H. As a consequence of the results of the previous section, we see that any stochastic hybrid process is a Borel right process. By product, using the fact that any stochastic hybrid process has embedded in its structure some diffusion processes and a jump process, we obtain the expression of the infinitesimal generator. This will constitute the base for any type of stochastic calculus from now on.

Strong Markov Property   Stochastic hybrid processes, being constructed as particular Markov strings, have realisations that inherit the properties of their diffusion components, namely they are *strong Markov processes* with the *càdlàg property*.

Proposition 4.9 (Strong Markov process) *Any stochastic hybrid process M_H is a strong Markov process.*

Corollary 4.10 *Any stochastic hybrid process M_H is a Borel right process.*

As we discussed in the context of Markov strings, a GSHS realisation might be thought of as a 'restriction' of a random evolution process [214], whose components are diffusion processes defined on different state spaces. We can suppose also that each diffusion evolves in a larger space $\overline{\mathbf{X}}$. The first difference is that the state space for a GSHS is only $\bigcup_{q \in Q} \{q\} \times X^q$; usually a random evolution process is defined on the entire product space $Q \times \overline{\mathbf{X}}$. The second difference is that whilst for a random evolution process jump times are governed by continuous multiplicative functionals, for a GSHS these might also be given by some right continuous multiplicative functionals (e.g., the boundary hitting times of modes are defined by such discontinuous functionals).

Moreover, usually stochastic hybrid processes are right processes, but they may or may not be standard, Hunt, or Feller processes. Due to the predictable jumps that certainly take place when the trajectories hit the boundary, stochastic hybrid processes may lose continuity properties of the stopping times or of the operators of the transition semigroup.

The Process Generator   Denote by $\mathscr{B}_b(\mathbf{X})$ the set of all bounded measurable functions $f : \mathbf{X} \to \mathbb{R}$. This is a Banach space under the norm

$$\|f\| = \sup_{\mathbf{x} \in \mathbf{X}} |f(\mathbf{x})|.$$

The transition semigroup $(\mathbf{P}_t)$ is characterised by its *strong infinitesimal generator*, which is the 'derivative' of $\mathbf{P}_t$ at $t = 0$. Let $D(L) \subset \mathscr{B}_b(\mathbf{X})$ be the set of functions f for which the following limit (in the norm $\| \cdot \|$) exists:

$$\lim_{t \searrow 0} \frac{1}{t}(P_t f - f)$$

and denote this limit Lf.

By the *martingale property* (Proposition 2.12), if for $f \in D(L)$ we define the real-valued process $(C_t^f)_{t \geq 0}$ by

$$C_t^f = f(\mathbf{x}_t) - f(\mathbf{x}_0) - \int_0^t Lf(\mathbf{x}_s)\,ds \tag{4.21}$$

then for any $\mathbf{x} \in \mathbf{X}$, the process $(C_t^f)_{t \geq 0}$ is a local martingale on $(\Omega, \mathscr{F}, \mathscr{F}_t, \mathbb{P}_{\mathbf{x}})$.

By extending $D(L)$ with the set of functions that satisfy (4.21), we get the notion of *extended generator* of the process. Let $(\widehat{L}, D(\widehat{L}))$ be the extended generator of the process $(\mathbf{x}_t)$.

Following [64], for $A \in \mathscr{B}(\overline{\mathbf{X}})$ we define the following counting processes p, p^*:

- $p(t, A)$: the number of jumps in A until the time t for the process $(\mathbf{x}_t)$,

$$p(t, A) := \sum_{k=1}^{\infty} I_{(t \geq T_k)} I_{(\mathbf{x}_{T_k} \in A)};$$

- $p^*(t)$: the number of jumps from the boundary of the process $(\mathbf{x}_t)$,

$$p^*(t) := \sum_{k=1}^{\infty} I_{(t \geq T_k)} I_{(\mathbf{x}_{T_k^-} \in \partial \mathbf{X})}.$$

Then we may define $\widetilde{p}(t, A)$ as the compensator of $p(t, A)$ (see [64] for further explanation):

$$\widetilde{p}(t, A) := \int_0^t R(\mathbf{x}_s, A)\lambda(\mathbf{x}_s)\,ds + \int_0^t R(A, \mathbf{x}_{s-})\,dp^*(s),$$

$$\widetilde{p}(t, A) = \sum_{T_k \leq t} R(\mathbf{x}_{T_k-}, A).$$

The process

$$q(t, A) := p(t, A) - \widetilde{p}(t, A)$$

is a local martingale.

Given a function $f \in \mathscr{C}^1(\mathbb{R}^n, \mathbb{R})$ and a vector field $b : \mathbb{R}^n \to \mathbb{R}^n$, we use $\mathscr{L}_b f$ to denote the Lie derivative of f along b given by

$$\mathcal{L}_b f(x) = \sum_{q=1}^{n} \frac{\partial f}{\partial x_q}(x) b_q(x).$$

Given a function $f \in \mathscr{C}^2(\mathbb{R}^n, \mathbb{R})$, we use $\mathbb{H}^f$ to denote the Hamiltonian operator applied to f, i.e.,

$$\mathbb{H}^f(x) = \big(h_{ij}(x)\big)_{q,j=1,\dots,n} \in \mathbb{R}^{n \times n},$$

where

$$h_{ij}(x) = \frac{\partial^2 f}{\partial x_q \partial x_j}(x).$$

A^{T} denotes the transpose matrix of a matrix $A = (a_{ij})_{q,j=1,\dots,n} \in \mathbb{R}^{n \times m}$ and $\mathrm{tr}(A)$ denotes its trace.

Theorem 4.11 (GSHS infinitesimal generator) [28] *Let H be a GSHS as in Definition 4.1. Then the domain $D(\widehat{L})$ of the extended generator $\widehat{L}$ of M_H, as a Markov process, consists of at least those measurable functions f on $\mathbf{X} \cup \partial \mathbf{X}$ satisfying the following.*

1. $f : \overline{\mathbf{X}} \to \mathbb{R}$ *is $\mathscr{B}$-measurable such that for each $q \in Q$ the restriction $f^q = f|_{X^q}$ is twice differentiable.*
2. *The boundary condition*

$$f(\mathbf{x}) = \int_{\mathbf{X}} f(\mathbf{y}) R(\mathbf{x}, d\mathbf{y}), \quad \mathbf{x} \in \partial \mathbf{X}. \tag{4.22}$$

3. $Bf \in L_1^{loc}(p),^2$ *where*

$$Bf(\mathbf{x}, s, \omega) := f(\mathbf{x}) - f\big(\mathbf{x}_{s-}(\omega)\big). \tag{4.23}$$

4. *For $f \in D(\widehat{L})$, $\widehat{L} f$ is given by*

$$\widehat{L} f(\mathbf{x}) = L_{cont} f(\mathbf{x}) + L_{jump} f(\mathbf{x}) \tag{4.24}$$

where

$$L_{cont} f(\mathbf{x}) := \mathscr{L}_b f(\mathbf{x}) + \frac{1}{2} \mathrm{tr}\big(\sigma(\mathbf{x})\sigma(\mathbf{x})^{\mathsf{T}} \mathbb{H}^f(\mathbf{x})\big), \tag{4.25}$$

$$L_{jump} f(\mathbf{x}) := \lambda(\mathbf{x}) \int_{\overline{\mathbf{X}}} \big(f(\mathbf{y}) - f(\mathbf{x})\big) R(\mathbf{x}, d\mathbf{y}). \tag{4.26}$$

For the proof of this theorem, we refer the reader to [28]. Related to this theorem, there are some aspects that need to be discussed. At the time of this research, we do not have the precise characterisation of the whole domain $D(\widehat{L})$ corresponding

---

[2]Following [64], f is in $L_1^{loc}(p)$ if for some sequence of stopping times $\sigma_n \uparrow \infty$

$$E_x \sum_i \big| f(x_{T_i \wedge \sigma_n}) - f(x_{T_i \wedge \sigma_n -}) \big| < \infty.$$

to an extended generator of a stochastic hybrid process. In the particular case of piecewise deterministic Markov processes, such a characterisation exists (see [64]), but this is due to the fact that the PDMPs are nondiffusion processes and we deal with fewer restrictions. For dealing with the extended generator $\widehat{L}$, it is enough to know, at least, a core of $D(\widehat{L})$ provided in the above theorem. As we can see, even to check the conditions (4.22), (4.23) might be a difficult task. Then working directly with the martingale problem might be a better alternative.

Regarding the expression of the generator (4.24), it is clear that it combines the generators for a continuous diffusion process (4.25) with the generator of a jump Markov process (4.26). Then (4.24) shows that a stochastic hybrid process behaves locally either as a diffusion process or a jump process. In other words, its dynamics is either continuous or discrete in an infinitesimal time interval.

4.5 Examples of Stochastic Hybrid Systems

In this section, we present a collection of hybrid system examples, which can be modelled in the GSHS framework.

4.5.1 Single-Server Queues

This example was presented in [65] as a model belonging to the class piecewise deterministic Markov processes, which is a particular class of GSHS.

Customers arrive at a queue at random times $T_1, T_2, \ldots$, and the customer arriving at time T_q requires $Y_q > 0$ units of time for processing. The total service load (in time units) presented up time t is

$$L_t = L_0 + \sum_q Y_q I_{(t \geq T_q)}$$

where $L_0 \geq 0$ is the service load existing at time 0, and the *virtual time* V_t is the unique solution of the equation

$$V_t = L_t - \int_0^t I_{[V_s > 0]} \, ds.$$

The first term on the right is the total service load presented whilst the second is the amount of carried out processing; i.e., V_t is the amount of unprocessed load at time t, or equivalently the time a customer arriving at t would have to wait for service to begin. The queue has two possible configurations, namely *busy* and *empty*, and we denote these by an indicator variable q with $q = 1$ when the queue is busy and $q = 0$ when it is empty.

The queueing systems are characterised via the conventional classification $A/B/n$, where A refers to the arrival process, B to the distribution of Y_q and n to the number of servers ($n = 1$ in this case). Let us consider the $M/G/1$ queue, where M (for 'Markov') means that the interarrival times are independent and exponential (i.e., the arrivals form a Poisson process), and G (for 'general') means that Y_q are independently identically distributed with some arbitrary distribution F. Y_q are assumed to be independent of the arrivals process.

When

- $q = 1$, V_t decreases at unit rate between jumps;
- $q = 0$ there is an exponential waiting time for a new arrival (because of memoryless property of the exponential distribution).

The state space of this model is

$$\mathbf{X} = \{0, 0\} \cup \{1\} \times (0, \infty)$$

and we take $\mathbf{x}_t = (q_t, V_t)$, where q_t is the current configuration of the queue and V_t is the virtual waiting time. Then $(\mathbf{x}_t)$ is a Markov process, which evolves as follows: when V_t hits zero, $\mathbf{x}_t$ jumps to $(0, 0)$ and waits there until the next arrival, at which point it jumps to $(1, Y)$, where Y is the service requirement of the arriving customer.

In the case of the $GI/G/1$ queue: GI (for 'general independent') means that the interarrival times are independently identically distributed with general distribution. In order to get a Markov process, it is necessary to include the supplementary variable τ, the time since the last arrival. The state space is

$$\mathbf{X} = \{0\} \times [0, \infty) \cup \{1\} \times [0, \infty) \times (0, \infty).$$

When $q = 1$, $\zeta = (\zeta^1, \zeta^2)$ is two dimensional with $\zeta_t^1 = \tau$, and $\zeta_t^2 = V_t$ the virtual waiting time. Denote $\mathbf{x}_t = (q_t, \zeta_t)$. When $q = 1$, ζ_t^1 and ζ_t^2 increase and decrease, respectively, at unit rate. If the queue becomes empty (i.e., $\zeta^2 = 0$) then it is necessary to continue accumulating the time since the last jump, so $\mathbf{x}_t$ jumps from $(1, \zeta^1, 0)$ to $(0, \zeta^1)$. If the next arrival occurs t time units later, bringing service requirement Y, then $\mathbf{x}_t$ jumps from $(0, \zeta^1 + t)$ to $(1, (0, Y))$.

4.5.2 A Hybrid Manufacturing System Model

This model has been studied in [107] and is motivated by the structure of many manufacturing systems. In these systems, discrete entities (referred to as *jobs*) move through a network of work centres, which process the jobs so as to change their physical characteristics according to certain specifications. Associated with each job is a *temporal state* and a *physical state*. The temporal state of a job evolves according to event-driven dynamics and includes information such as arrival time, waiting time, service time or departure time of the job at the various work centres. The physical state evolves according to time-driven dynamics modelled through differential equations, which, depending on the particular problem that is studied, describe changes in such quantities as the temperature, size, weight, chemical composition, bacteria level, or some other measures of the 'quality' of the job. The interaction of time-driven with event-driven dynamics leads to a natural trade-off between temporal requirements on job completion times and physical requirements on the quality of the completed jobs.

Consider a single-stage manufacturing process modelled as a single-server queueing system. The server processes one job at a time on a first-come first-served nonpreemptive basis (i.e., once a job begins service, the server cannot be interrupted, and will continue to work until the operation is completed). Identical jobs arriving at the system with rate λ wait in an infinite capacity queue until they are processed by the server operating at rate $\lambda \in \Lambda$. *Exponential interarrival times are assumed.* The controller is assumed to select processing rates from a finite set

$$\Lambda = \{\lambda_1, \lambda_2, \ldots, \lambda_m\} \quad \text{where } \lambda_q < \lambda_{q+1}, \ q = 1, \ldots, m - 1.$$

As job q is being processed, its physical state, denoted by x_q, evolves according to time-driven dynamics of the general form

$$\dot{x}_q = g_q\left(x_q, \lambda_q(t), t, \omega\right), \qquad x_q(\tau_q) = x_q^0, \tag{4.27}$$

where τ_q is the time when processing begins and x_q^0 is the initial state at that time. The control variable $\lambda_q(t)$ is used to attain a final desired physical state corresponding to a target 'quality level'. On the other hand, the temporal state of the qth job is denoted by T_q^e and represents the time when the job completes processing and departs from the system. Letting T_q be the arrival time of the qth job and S_q be the service time which is a function of $\lambda_q(t)$ during the process, the event-driven dynamics describing the evolution of the temporal state is given by the following 'max-plus' recursive equation:

$$T_q^e = \tau_q + S_q = \max\left(T_q, T_{q-1}^e\right) + S_q, \tag{4.28}$$

where $T_0^e = -\infty$ in which case $\tau_1 = T_1$ and the first job begins service as soon as it arrives. Equation (4.28) is known in queueing theory as the Lindley equation.

The system is hybrid in the sense that it combines the time-driven dynamics (4.27) with the event-driven dynamics (4.28). If we suppose that there are no delays, i.e., $T_{q-1}^e = T_q$ for $q = 2, \ldots, m$ then this model is a particular case of GSHS with no forced discrete transitions.

4.5.3 A Simplified Model of a Truck with Flexible Transmission

This example was used in [112, 145]. The system is described by

$$dx_1 = x_2 \, dt,$$
$$dx_2 = -x_2 + x_3,$$
$$dx_3 = -x_2 + g_q(x_2)u \, dt + \sigma \, dw,$$
$$q = 1, 2, \quad -0.1 \le u \le 1.1, \quad \sigma = 0.01,$$

where x_1, x_2 and x_3 are the position, velocity and the rotational displacement of its transmission shaft, respectively. The efficiency for gear q is $g_q(x)$, u is the throttle, and dw is a scalar Wiener process. In [145], the model is modified in comparison with the one presented in [112] by assuming that gear switches occur at the speed of equal efficiency between the gears ($x_2 = 0.5$) and, therefore, the switching boundary is defined by

$$A = \{x \,|\, x_2 = 0.5\}.$$

The objective is to drive the system from the state (x_0, q_0) to the target set

$$E = \left\{ x \left| \frac{1}{2} x^T x \le 0.25 \right. \right\}.$$

4.5.4 The Stochastic Thermostat

This is an example of a GSHS with two discrete states, which models the temperature in a house with n rooms, $n \ge 1$, regulated by a single thermostat. This is the generalisation of the one-dimensional process that was studied in [178].

Let $z = (z_1, z_2, \ldots, z_3) \in \mathbb{R}^n$ describe the temperature in the n rooms of the house, and $q \in Q = \{0, 1\}$ the binary state of the thermostat. The global state of the system is then described by the hybrid state $x = (q, z) \in Q \times \mathbb{R}^n$, which has both a discrete and a continuous component. For a given discrete state $q \in Q$ of the thermostat, the temperature z_t evolves in $\mathbb{R}^n$ according to an SDE of the form

$$dz_t = b(q, z_t) + \sigma \, dW_t, \tag{4.29}$$

where $\sigma \in \mathbb{R}^{n \times n}$ and $b_q = b(q, \cdot)$ describe the action of the thermostat, the effect of the exterior environment, and the coupling between the temperatures of adjacent rooms. The switching of the thermostat is controlled by a linear criterion $\Psi(z) = \sum_{q=1}^{n} \alpha_q z_q$. The thermostat switches on when $\Psi(z_t)$ crosses some threshold $\Psi_{\min}$ downwards and switches off when it crosses another threshold $\Psi_{\max} > \Psi_{\min}$ upwards. This can be described in the GSHS framework as follows: We define

$$X_0 = \left\{ z \in \mathbb{R}^n \,\middle|\, \Psi(z) > \Psi_{\min} \right\},$$
$$X_1 = \left\{ z \in \mathbb{R}^n \,\middle|\, \Psi(z) < \Psi_{\max} \right\}$$

and then the hybrid state space is

$$\overline{\mathbf{X}} = \{0\} \times \overline{X}_0 \cup \{1\} \times \overline{X}_1 \subset Q \times \mathbb{R}^n.$$

The process z_t evolves continuously in X_q according to the SDE (4.29) as long as the thermostat is in state $q_t = q$, and Q_t switches when z_t reaches ∂X_q. Therefore, the hybrid process $\mathbf{x}_t = (q_t, z_t)$ is a GSHS with only forced discrete transitions.

4.6 Some Remarks

In this chapter, we formally set up the concepts of stochastic hybrid systems and stochastic hybrid processes. Stochastic hybrid systems are presented in the hybrid automata formalism and the stochastic hybrid processes represent their dynamics. The latter ones can be described using the notion of a Markov string, which is, roughly speaking, a concatenation of Markov processes. This notion has arisen as a result of our research on stochastic hybrid system modelling [28, 42, 43, 193] and it aims to be a very general formalisation of all existing models of stochastic hybrid processes. The Markov string concept has proven to be a very powerful tool in the studying of the general models of stochastic hybrid systems introduced at the beginning of the chapter.

One of the main contributions of this work is the proof of the strong Markov property. Since stochastic hybrid processes constitute a particular class of Markov strings, this property holds also for them.

At the end of this chapter, based on the strong Markov property of these processes, we derived the expression of the infinitesimal generator that characterises the Markov evolution of a GSHS. The connection with the martingale problem is also discussed. Very important for the well-posedness of this problem is the càdlàg property of hybrid trajectories. These developments constitute the basis of the stochastic calculus that can be handled for GSHS.

The strong Markov is essential for defining stochastic control problems, like the stochastic reachability, which is the subject of this monograph. The expression of the infinitesimal generator is the most illustrative mathematical object able to capture the hybridicity of GSHS.

This chapter has a structure similar to the chapter entitled 'Stochastic Hybrid Systems: Theory and Safety Critical Applications' in [28]. In comparison to [28], we have simplified the proof for the strong Markov property, and the construction of Markov strings has been explained in more detail, following a step-by-step algorithm with illustrating figures. The martingale problem for stochastic hybrid process has only been discussed and no proofs provided. The reason for this is the fact that the extended generator domain, and the well-posedness of the martingale problem for stochastic hybrid processes are still open research problems.

Chapter 5
Stochastic Reachability Concepts

5.1 Overview

This chapter is intended to be a background chapter with respect to our main topic, namely, stochastic reachability. Stochastic reachability for hybrid systems is a theme that belongs to a larger research area called *stochastic hybrid control*. After a short introduction in the topics that are at the core of the research on stochastic hybrid control, we present the concept of stochastic reachability. Due to the fact that the trajectories of a random process (hybrid or not hybrid) are not uniquely defined by the initial state, the reachability problem in stochastic framework consists of finding an appropriate measure for the process trajectories that start with a given initial state and visit a target set. The first step in the research effort of stochastic reachability is proving that the concept is well defined. We do this using an argument based on analytic sets. The next step is relating the stochastic reachability to other concepts associated with the underlying stochastic process in order to smooth the way that leads to the estimation of reachability measures.

5.2 Stochastic Hybrid Control

Control problems are at the core of the research in stochastic hybrid systems. This is because many important applications with prominent hybrid stochastic dynamics come from the area of cyber-physical system control. For example, stochastic hybrid control has played an important role in applications to aerospace (unmanned aerial vehicles, air traffic management, satellites), industrial process control (biodiesel production) and manufacturing (paper making) and robotics (balance and falling).

First, the models for control problems that have arisen in these applications differ from one another in how they exhibit randomness. Stochastic hybrid systems may have random discrete events (and timing) or random continuous dynamics or both. Their behaviour requires the most general models of dynamical systems in the sense that time-driven stochastic systems and general nonlinear systems can (technically) all be viewed as special cases of stochastic hybrid systems. To model the

L.M. Bujorianu, *Stochastic Reachability Analysis of Hybrid Systems*,
Communications and Control Engineering,
DOI 10.1007/978-1-4471-2795-6_5, © Springer-Verlag London Limited 2012

randomness inherent to discrete events we have to consider that discrete transitions are governed by a controlled Markov chain. If this is the case, a probability distribution is assumed to model disturbance input. This extra information can be exploited by the discrete/continuous controller and, it gives us the option to formulate finer requirements as well. For example, we can ask that some specifications be satisfied with almost sure probability one on a long run of the controlled Markov chain (ergodicity problem) or that some properties (until a certain event appears) have probability less than a given threshold (viability problem).

For stochastic hybrid systems, classical control problems (studied in the literature) need to be specified in a different way. Generally, according to the specification, the control problems can also be grouped into three classes:

(1) *Stabilisation.* The stability problem for hybrid systems aims to select the continuous inputs and/or the timing and destinations of discrete transitions to ensure that the system remains close to an equilibrium point, limit cycle or other invariant set. Variants of this problem abound in the literature. They consider different assumptions regarding

- control inputs (discrete/continuous with both discrete/continuous probability distributions) and
- type of stability specification (stabilisation, asymptotic/exponential stabilisation, stability in distribution, moment stability, almost sure asymptotic stability, etc.).

(2) *Stochastic optimal control.* The optimal control problem for hybrid systems consists of finding the optimal (continuous and/or discrete) controls that steer the hybrid system to a specific evolution such that a certain cost function is minimised. Again, different variants have been considered in the literature, according to

- the types of available (discrete/continuous) input,
- the accumulated cost one can consider during the hybrid evolution (along the continuous part or discrete part or both) and
- the time horizon over which the optimisation is carried out, which might be finite or infinite or another type.

(3) *Language specifications.* These interesting control problems are specified with respect to the hybrid trajectories. Such problems require that the trajectories of the hybrid system be contained in a set of desirable trajectories with a certain probability measure. Usually, trajectory constraints arise from reachability considerations, of either

- the *safety* type (along all trajectories, the state of the system should be kept in a good region of the state space) or
- the *liveness* type (along all trajectories, the system state may eventually reach a good region of the state space).

Starting with these simple primitives, more complex specifications can be formulated such as:

- the system should visit a given set of states infinitely often,
- given two sets of states, if the system visits one infinitely often it should also visit the other infinitely often, and
- given two sets of states, if the system visits one set then it will not visit the other set.

These specifications have been to a large extent motivated by analogous problems specified using temporal logics formulated for discrete systems.

Moreover, such problems can be combined with optimal control problems giving rise to even more complex specifications like:

- minimise the cost that the system will visit a given set of states infinitely often,
- minimise the cost that the system trajectories will remain in a certain set,
- maximise the cost that the system trajectories will hit a target set, etc.

5.3 Stochastic Reachability Problem

In many situations, the modelling paradigms used for stochastic hybrid systems are rather complex (involving nonlinear dynamics, continuous probability distributions) and therefore the goal of finding finite abstractions or approximations is not always reasonable. Stochastic analysis tools have been proposed [42, 43] as an alternative verification method for situations when uncertain or probabilistic systems do not admit finite state abstractions. Although computer science communities, understandably, prefer discrete probabilistic models, other branches of engineering employ the widespread use of many more complex stochastic models. The most important verification method for these models is reachability analysis. The stochastic features of these models have to be captured in the formulation of the reachability problem. This leads to the concept of *stochastic reachability*. This problem is usually formulated as a control problem with respect to the hybrid trajectories.

5.3.1 Motivation

Understanding the reachability problem is among the most important objectives in the context of a stochastic hybrid modelling.

In general terms, a reachability problem consists of determining if a given system trajectory will eventually enter a prespecified set starting from some initial state. For deterministic hybrid systems, reachability analysis refers to the problem of computing bounds on the set of states that can be reached by system trajectories. Reachability analysis is relevant to a variety of control applications. The available reachability tools consider various uncertainty parameters. In this context 'uncertainty' is synonymous with nondeterminism. Nondeterminism descriptions might have different mathematical incarnations such as differential inclusions [198, 223], polygonal approximations [55], ellipsoidal approximations [156] and general nonlinear systems with set disturbances [170].

In many control applications, the dynamics of the system under study is sub-jected to the perturbation of random noises that are either inherent or present in the environment. Typically, a certain part of the state space is 'unsafe' and the control input to the system has to be chosen so as to keep the state away from it, despite the presence of random noises. This can happen in many safety-critical situations. Therefore, in these applications it is very important that one has some measure of criticality for evaluating whether the selected control input is appropriate or a cor-rective action should be taken to steer the system out of the unsafe set in a timely way. A natural choice for the measure of criticality is the probability of intrusion into the unsafe set within a finite/infinite time horizon: the higher the intrusion prob-ability, the more critical is the situation. Within the air traffic management (ATM) context, safety-critical situations arise during flight when an aircraft comes closer than a minimum allowed distance to another aircraft or enters a forbidden region of the airspace. In the current ATM system, air traffic controllers are in charge of guaranteeing safety by issuing to pilots corrective actions on their flight plans when a safety-critical situation is predicted. The limit on air traffic system capacity ow-ing to its human-operated structure can be pushed forward by introducing automatic tools that support air traffic controllers in detecting safety-critical situations.

For *safety-critical air traffic management situations*, one important goal was to develop formal mathematical models [42, 43, 193]. A central problem in air traffic control is determining the *collision probability* (rare events [18]), i.e., the probability that two aircraft come closer than a minimum allowed distance. If this probability can be computed, an alert can be issued when it exceeds a certain threshold.

In the context of stochastic hybrid systems, the computation of conflict probabil-ity reduces to a *reachability problem*: *computing the probability that the stochastic hybrid process modelling the aircraft motion reaches an unsafe part of the state space* (where two aircraft come closer than the minimum allowed distance). This is the key approach in new air traffic control philosophies like the *free flight concept* [119]. Free flight, sometimes referred to as *self separation assurance* [119], is a con-cept where pilots are allowed to select their trajectory freely in real time, at the cost of acquiring responsibility for conflict prevention. It changes air traffic management in such a fundamental way that one could speak of a paradigm shift: the centralised control becomes a distributed one, responsibilities transfer from ground to air, air traffic sectorisation and routes are removed and new technologies are brought in. In the RTCA Free Flight Task Force, free flight is presented as a range of concepts, allowing self optimisation of the routes by the airlines. The document also describes a mechanism for airborne separation as a part of the free flight concept. The basic assumption is that the collision probability becomes smaller as the sky gets less crowded. Safety analysis of free flight is essentially stochastic. The Achilles heel of the concept is then that reachability analysis needs to be stochastic.

Stochastic hybrid system models developed in this book could be applied to the aircraft dynamics model, which incorporates information on the aircraft flight plan and takes into account adversarial weather conditions as the main source of uncer-tainty for the actual aircraft motion. Therefore, the problem of estimation of prob-ability that an aircraft enters an 'unsafe' region can be formally specified as the stochastic reachability problem for stochastic hybrid systems.

5.3.2 *Mathematical Formulation*

The Markov property holds for almost all hybrid system models discussed in this monograph, in particular for a GSHS, the model presented in detail in Chap. 4. Because of this, we are allowed to formulate the reachability problem in the general setting of Markov processes. Assume that we have a given strong Markov process,

$$\mathbf{M} = (x_t, \mathbb{P}_x).$$

We further assume that (x_t) has the state space $(\mathbf{X}, \mathscr{B})$ (where $\mathbf{X}$ is a Borel space and $\mathscr{B}$ is the Borel σ-algebra of $\mathbf{X}$) and it has the càdlàg property (its paths are right continuous with left limits).

Let A be a Borel set of the state space $\mathbf{X}$, i.e., $A \in \mathscr{B}(\mathbf{X})$. In this context the trajectories of the system can be identified with the elementary events of the underlying probability space Ω. Thus, the reachable 'event' associated to A (i.e., the set of trajectories that reach A within a finite/infinite time horizon) can be defined as follows:

$$Reach_T(A) = \{\omega \in \Omega \mid \exists t \in [0, T] : x_t(\omega) \in A\}, \tag{5.1}$$

$$Reach_\infty(A) = \{\omega \in \Omega \mid \exists t \geq 0 : x_t(\omega) \in A\} \tag{5.2}$$

(keep in mind that an element $\omega \in \Omega$ is, in fact, a trajectory of the system).

Developing a methodology for the reachability analysis of SHS will involve dealing with two aspects:

1. the theoretical aspect of the measurability of reachability sets (the set of trajectories that enter a prespecified set of state space);
2. the computational aspect regarding how to estimate the probability of reachable events and how to quantify the level of approximation introduced.

Problem 1 We would like to have some information about the measure $\mathbb{P}[Reach_T(A)]$ and $\mathbb{P}[Reach_\infty(A)]$ of these sets in the underlying probability space. But for this we should know if we are allowed to apply the measure $\mathbb{P}$ to the sets defined by (5.1) and (5.2). Therefore, the first problem can be formulated as follows:

- Are $Reach_T(A)$ and $Reach_\infty(A)$ really events?

Remark 5.1 Note that if $Reach_T(A)$ is an event then $Reach_\infty(A)$ is also an event, since

$$Reach_\infty(A) = \bigcup_{n=0}^{\infty} Reach_n(A).$$

Problem 2 If it turns out that we can assign a probability to $Reach_T(A)$ and $Reach_\infty(A)$, the next problem can be formulated as follows:

- Can we compute these probabilities?

The next section will solve the measurability issues of reachable events. Then, in the following sections the stochastic reachability probabilities will be characterised in terms of other probabilistic objects from Markov processes like hitting/exit distributions, occupation measures, upper or lower probabilities. These characterisations will be used further in this book to obtain appropriate evaluations of the reach probabilities.

5.4 Measurability Results

The measurability of the reachable events can be proved employing some standard techniques for Markov processes using the properties of the so-called *analytic sets* (see the Appendix). The method is very general and it works for all stochastic hybrid system models, provided they are Markov processes with the càdlàg property.
  Let

$$\mathbf{M} = (x_t, \mathscr{F}, \mathscr{F}_t, \mathbb{P}_x)$$

be a Markov process with the càdlàg property with state space $\mathbf{X}$. Suppose $\mathbf{X}$ is a Borel space. According to [183], we can choose Ω to be $D_{\mathbf{X}}[0, \infty)$, i.e., the set of right continuous functions with left limits with values in $\mathbf{X}$. $D_{\mathbf{X}}[0, \infty)$ is a Borel space.
  The process $\mathbf{M}$ is called *measurable* if the function

$$x : (t, \omega) \rightarrow x_t(\omega)$$

is measurable on the product space $(\mathbb{R}_+ \times \Omega, \mathscr{B}(\mathbb{R}_+) \otimes \mathscr{F})$, i.e.,

$$x \in \mathscr{B}(\mathbb{R}_+) \otimes \mathscr{F} / \mathscr{B}(\mathbf{X}).$$

Proposition 5.2 [83] *All Markov processes with the càdlàg property are measurable processes.*

Theorem 5.3 *Let $A \in \mathscr{B}(\mathbf{X})$ be a given Borel set. Then $Reach_T A$ and $Reach_\infty(A)$ are universally measurable sets in Ω.*

Proof Since the process has the càdlàg property, $x : \mathbb{R}_+ \times \Omega \rightarrow X$ is a $\mathscr{B}(\mathbb{R}_+) \times \mathscr{F}^0 / \mathscr{B}(X)$-measurable function (see Proposition 5.2). Since $A \in \mathscr{B}(X)$, clearly $x^{-1}(A) \in \mathscr{B}(\mathbb{R}_+) \times \mathscr{F}^0$. For $T > 0$, we set

$$A_T(A) = x^{-1}(A) \cap [0, T] \times \Omega.$$

Since $[0, T] \times \Omega \in \mathscr{B}(\mathbb{R}_+) \times \mathscr{F}^0$ then $A_T(A) \in \mathscr{B}(\mathbb{R}_+) \times \mathscr{F}^0$. Now, using Theorem A.3 (from the Appendix), we obtain

$$Reach_T(A) = Proj_\Omega A_T(A)$$

is an $\mathscr{F}^0$-analytic set. Therefore, because Ω is a Borel space, this implies (cf. Proposition A.5) that $Reach_T(A)$ is a universally measurable set. Obviously, $Reach_\infty(A)$

Fig. 5.1 Reachability
connections

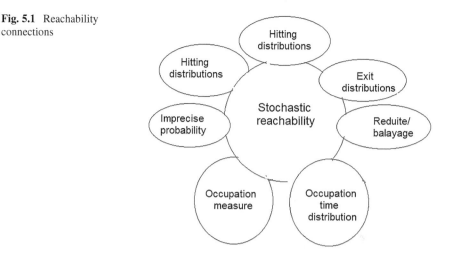

is, also, a universally measurable set. Since $\mathcal{U}_\Omega$, the σ-measurable sets, is included in $\mathcal{F}$ (which is the completion of $\mathcal{F}^0$ with respect to all probabilities P^μ, where μ runs in the set of all probability measures on X), $Reach_T(A)$ belongs to $\mathcal{F}$.   □

Remark 5.4 Theorem 5.3 is very general; the proof requires the Markov process to be measurable and its probability space to be a Borel space.

5.5 Different Measures Associated with Stochastic Reachability

In this section, we relate the concept of stochastic reachability with other concepts associated to the underlying Markov process. These concepts provide useful insights for studying different facets of the stochastic reachability problem.

Briefly, these concepts (see Fig. 5.1) are:

- *Hitting distributions*: These capture the probability of the random variable that describe when the system trajectories hit, for the first time, the target set.
- *Exit distributions*: These capture the probability of the random variable that describe when the system trajectories exit, for the first time, a 'good' set of the state space.
- *Occupation measure*: This captures the mean of the random variable that 'record' what part of the state space has been swept by the process until hitting a 'bad' set.
- *Occupation time distributions*: These capture the 'speed' of the time spent in a certain set.
- *Imprecise probabilities*: These regard the hitting probabilities as set functions that apply to the target sets.
- *Réduite and balayage*: These are concepts of potential theory that are useful in the characterisation of the reach probabilities as solutions of some appropriate boundary value problems.

5.5.1 Hitting Distributions

The probabilities of reach events (5.1) and (5.2) are, respectively,

$$\mathbb{P}(T_A < T) \quad \text{and} \quad \mathbb{P}(T_A < \zeta), \tag{5.3}$$

where ζ is the lifetime of $\mathbf{M}$ and T_A is the first hitting (passage) time of A

$$T_A = \inf\{t > 0 | x_t \in A\} \tag{5.4}$$

and $\mathbb{P}$ is a probability on the measurable space $(\Omega, \mathscr{F})$ of the elementary events associated to $\mathbf{M}$. $\mathbb{P}$ can be chosen to be $\mathbb{P}_x$ (if we want to consider the trajectories that start in x) or $\mathbb{P}_{\mu_0}$ (if we want to consider the trajectories with an initial condition chosen according to an initial probability distribution μ_0).

On the Borel σ-algebra $\mathscr{B}(\mathbf{X})$, we can define a new measure called *hitting distribution* by

$$H^A(B) := \mathbb{P}(x_{T_A} \in B), \tag{5.5}$$

where T_A is the first hitting time of A given by (5.4). The quantity $H^A(B)$ assesses the probability that the first entry time in A belongs B. In other words, this measure helps us to make a refinement of the target set A and to find which parts of A are most likely to be hit first by process paths.

If in the expression (5.5) we replace $\mathbb{P}$ by the Wiener probabilities $\mathbb{P}_x$, we can consider only the trajectories that start in x, i.e.,

$$H^A(x, B) := \mathbb{P}_x(x_{T_A} \in B) = \mathbb{E}_x 1_B(x_{T_A}),$$

where $\mathbb{E}_x$ is the expectation with respect to $\mathbb{P}_x$ and 1_B is the indicator function of B.

Therefore, extending this quantity from measurable sets to measurable functions, we may define the *hitting operator*

$$P_A : \mathscr{B}^b(\mathbf{X}) \to \mathscr{B}^b(\mathbf{X})$$

associated to the underlying Markov process (x_t):

$$P_A v(x) := \mathbb{E}_x\{v \circ x_{T_A} | T_A < \zeta\}. \tag{5.6}$$

The concept of hitting distributions corresponding to the target set A is used in connection with the restriction of the given Markov process (x_t) until the first entry time in A. One can define a new Markovian transition function

$$p_t^0(x, E) := \mathbb{P}_x(x_t \in E | t < T_A)$$

for $x \in \mathbf{X}$, $E \in \mathscr{B}(\mathbf{X})$. The restriction of $\{p_t^0, t > 0\}$ to $(\mathbf{X} \backslash A, \mathscr{B}(\mathbf{X} \backslash A))$ is the transition function of the part $\mathbf{M}_{\mathbf{X} \backslash A}$ of the process $\mathbf{M}$ on $\mathbf{X} \backslash A$. In this context, we can define the α-order hitting distributions by

$$H_\alpha^A(x, B) := \mathbb{E}_x\left[e^{-\alpha t} \cdot 1_B(x_{T_A})\right]$$

for $\alpha > 0$, $x \in \mathbf{X}$, $B \in \mathscr{B}(\mathbf{X})$. For a Borel measurable set A, H_α^A is a stochastic kernel on $(\mathbf{X}, \mathscr{B}(\mathbf{X}))$. Moreover, using the strong Markov property of $\mathbf{M}$, a variant of the *Dynkin formula* can be proved:

$$V_\alpha f(x) = V_\alpha^0 f(x) + H_\alpha^A f(x), \quad x \in \mathbf{X}, \ f \in \mathbf{B}^+(\mathbf{X}). \tag{5.7}$$

In this formula, $(V_\alpha)_{\alpha>0}$ (respectively $(V_\alpha^0)_{\alpha>0}$) represents the resolvent (i.e., the Laplace transform) corresponding to the transition probabilities $\{p_t, t > 0\}$ (respectively $\{p_t^0, t > 0\}$). For example, the resolvent $(V_\alpha^0)_{\alpha>0}$ for $\mathbf{M}_{\mathbf{X}\backslash A}$ is given by

$$V_\alpha^0(x, E) := \int_0^\infty e^{-\alpha t} \cdot p_t^0(x, E) \, dt$$

$$= \mathbb{E}_x \left[\int_0^{T_A} e^{-\alpha t} \cdot I_E(x_t) \, dt \right].$$

If u is α-excessive, then so is its α-*harmonic average* $H_\alpha^A u$ relative to the measurable set A. We set

$$p_A^\alpha(x) := H_\alpha^A 1(x) = \mathbb{E}_x \left(e^{-\alpha T_A} \right),$$

$$p_A(x) := H^A 1(x) = \mathbb{P}_x (T_A < \zeta).$$

Clearly, the reach probability for infinite horizon time can be expressed using these hitting distributions as follows.

Proposition 5.5 *For any $x \in \mathbf{X}$ and Borel set $A \in \mathscr{B}(\mathbf{X})$, we have*

$$\mathbb{P}_x \big[Reach_\infty(A) \big] = p_A(x) = \mathbb{P}_x [T_A < \zeta].$$

Keep in mind that p_A is still an excessive function. This fact will be used in the next chapters when we make the connection between reach probabilities and optimal stopping.

Here, we have to discuss also the reachability problem with finite horizon time. A fixed horizon time T is not a stopping time for the underlying stochastic process. Therefore, we cannot use the restriction of the given process until this time, since the Markov property might be lost. For stochastic hybrid systems, we may use the construction developed in Chap. 4 for killing the process after time T. In this way, the lifetime of process ζ will be replaced by T and all above discussions remain true. Then, in an unexpected way, it is easier to study the reachability problem with infinite horizon time than with finite horizon time.

Computational methods for reach probabilities in the form of first hitting time distributions still represent an open problem. The difficulties arise even in the case of simple diffusion processes. The problem of finding the probability distribution of the first hitting times of diffusion processes has been an important and difficult endeavour in stochastic calculus. It has applications in many fields of physics (first exit time of a particle in a noisy force field) or in biology and neuroscience (spike time distribution of a neuron with exponentially decaying synaptic current).

Even for one-dimensional diffusion processes, first passage time problems through a time-dependent boundary represent a challenging problem and have received a lot of attention over the last three decades. Unfortunately, the evaluation of the first passage time probability distribution function through a boundary (which might be fixed or time-dependent) is in general an arduous task and there are still no satisfactory solutions. Analytical results are scarce and fragmentary and depend on particular hypotheses. Since the problem is not solvable analytically, what remains is to study the asymptotic behaviour of this function and of its moments (see, e.g., [189]) or to the use ad hoc numerical procedures yielding approximate evaluations of the first passage time distributions.

Such approaches can be classified as follows: (i) those that are based on probabilistic methodologies (see, e.g., [75, 187]) and (ii) purely numerical methods, such as the widely used Monte Carlo method, which applies without any restriction, but whose results are rather coarse (for numerical methods, see, e.g., [143]).

In two and higher dimensions, the problem is even more complex and results can hardly be found. For particular types of Markov process the first hitting time distributions can be numerically evaluated through solving a backward Kolmogorov partial differential equation (PDE) with Dirichlet type boundary conditions. However, the numerical evaluation of such PDEs is limited to two- or, at most, three-dimensional problems.

For a special class of stochastic hybrid systems, namely for switching diffusion processes, approximations and Monte Carlo simulations for first passage time distributions have been developed in [148, 149]. Direct analytic solutions for this problem are hard to imagine. The viewpoint of this monograph on this subject is that we need, by any means, to abstract/approximate the underlying stochastic hybrid system in order to obtain sufficiently good evaluations for the first hitting probabilities.

5.5.2 Exit Distributions

Note that the first hitting time of A is equal to the first exit time from the complementary set of A,

$$E := A^c = \mathbf{X} \backslash A.$$

Then, the stochastic reachability problem can be formulated as an exit problem for the Markov processes that appear as realisations of stochastic hybrid systems.

Stochastic hybrid processes may be viewed as piecewise continuous jump diffusions, where the jumps are allowed to be spontaneous or forced (predictable).

If we look only to (continuous) pure diffusions processes, it is sufficient to consider the time when the process hits the boundary of E or A. However, when the stochastic processes also include jumps, then it is possible that the process overshoots the boundary and ends up in the exterior of the domain E (i.e., in the interior of A).

An important quantity that can be computed without explicit construction of the transition probability density function is the mean first passage time of the process from a specified domain. The mean first passage time is a measure of the stochastic time scale necessary for the process to be in a specified domain. Let E be a bounded subset of $\mathbb{R}^n$ with smooth boundary. Given $x \in E$, we want to know how long it takes for process x_t starting in x to leave domain E for the first time. We use

$$\tau_E^x := \inf\{t \geq 0 | x_0 = x, x_t \notin E\}.$$

The expectation of this random variable is called the *mean first passage time* MFPT of the first exit time:

$$\tau(x) := \mathbb{E}\tau_E^x.$$

For diffusion processes, we can calculate MFPT by solving an appropriate boundary value problem.

Theorem 5.6 *The MFPT is the solution of the boundary value problem*

$$\begin{cases} L\tau = -1, & x \in E \\ \tau = 0, & x \in \partial E \end{cases} \tag{5.8}$$

where L is the infinitesimal generator of the given diffusion process.

MFPT represents a well-studied problem in physics and biology [147] and in some specific network dynamics. The extension of a boundary value problem (5.8) to diffusions with jumps or to processes with predictable jumps (like the stochastic hybrid processes) reveal subtle difficulties due to the presence of these jumps. Moreover, in the end this effort would not be much appreciated, since MFPT illustrates an average of the probabilities defined in the context of stochastic reachability.

The difference between hitting probability distributions and MFPT is rather subtle. By definition, the MFPT is obtained by observing the first hitting times in an ensemble of trajectories represented by a given stochastic process and averaging those hitting times demonstrated by elements of this ensemble.

Usually, first passage (hitting) time distributions are computed using the *Fokker–Planck–Kolmogorov* (or *forward Kolmogorov*) *equations* associated to the underlying stochastic process. For stochastic hybrid systems, such equations were recently developed in [16, 17].

5.5.3 Occupation Measure

Another useful and more intuitive measure for stochastic reachability is the so-called *occupation measure*. Suppose that A is the target set and T_A is its first hitting time. We define the *occupation measure* μ_0 by

$$\mu_0(B) = \mathbb{E}\left[\int_0^{T_A} 1_B(x_s)\,ds\right]$$

for all $B \in \mathscr{B}(\mathbf{X})$, where 1_B is the indicator function of B. This measure evaluates what the probability is that a measurable set B is swept by the process (x_t) before hitting A. This measure allows us to introduce, for a given measurable bounded function positive f, a *reachability cost*, as follows:

$$J_x := \mathbb{E}_x \left[\int_0^{T_A} f(x_s) \, ds \right]. \tag{5.9}$$

If f depends also on some control inputs, then the goal of stochastic reachability will be to find the optimal control input that minimises the cost J_x.

When the process trajectories are continuous, the reachability cost (5.9) can be characterised analytically as a solution of some Hamilton–Jacobi equations. When the process trajectories are discontinuous, then we need to consider also the cost of jumps. In both cases, we have to use appropriate expressions for the infinitesimal generator of the process killed when it enters the target set A.

5.5.4 Occupation Time Distribution

For a sample path ω, we define the *occupation time distribution* up to the time $t > 0$ as follows:

$$L_t(\omega, B) := \frac{1}{t} \int_0^t 1_B(x_s(\omega)) \, ds, \quad B \in \mathscr{B}(\mathbf{X}). \tag{5.10}$$

If the process is conservative, then $L_t(\omega, \cdot)$ is a probability measure on $(\mathbf{X}, \mathscr{B})$. $L_t(\omega, B)$ represents the proportion of time the process spends in the set B during the time interval $[0, t]$.

Remark 5.7 One way to deal with stochastic reachability for target set A is to find how the expectation of the empirical measure defined by (5.10) corresponding to the complementary set E of A differs from the distribution of E.

The expectation of the occupation time distribution (viewed as a random variable)

$$\mathbb{E}_x[L_t(\omega, f)] := \frac{1}{t} \int_0^t \mathbb{E}_x f(x_s) \, ds, \quad f \in \mathscr{B}^b(\mathbf{X})$$

can be thought of as an *ergodic occupation measure* or '*long-run average*' cost when t is going to infinity. This measures the rare event probability, which is, in fact, the large deviation of $\int_0^t 1_B(x_s(\omega))$ with respect to its typical behaviour. Large deviation of occupation measures L_t for Markov processes is a traditional subject in probability, initiated by Donsker and Varadhan [69–72]. For stochastic reachability, the methods originated in large deviation theory can be applied when we deal with events whose probabilities are very small, i.e., *rare events*.

5.5.5 Imprecise Probability

Classical probability measures and linear mathematical expectations represent powerful tools for dealing with stochastic phenomena. However, in many applied areas, there are uncertain phenomena that cannot be easily modelled by additive measures and linear expectations.

We have seen that the characterisation of reachability measures employs concepts like hitting probabilities, excessive functions, occupation measures and so on. Let us consider the set function $v : \mathscr{B} \to [0, 1]$ defined by

$$v(A) := \mathbb{P}[Reach(A)]$$

with reachability defined in finite or infinite horizon time. This is a 'measure' of the target set. The natural problem is to find the appropriate mathematical concept that can capture such a measure. Clearly, though v is an increasingly monotone set function, v is *not* additive. Moreover, v is a subadditive measure, but it satisfies the following axiom:

$$v(A) = 1 - v(A^c)$$

where A^c is the complement of A.

Then the idea of using uncertainty measures for characterising how the trajectories hit a set seems to be a natural one. We will see that the stochastic reachability concept can be expressed in terms of imprecise probabilities. More precisely, considering the properties of the underlying stochastic hybrid process, the reach set measure v defines a *Choquet capacity* (in the way described in [51, 176]).

It might be useful to make a quick detour on nonadditive measures and nonlinear expectations with their applications. In 1953, Gustave Choquet introduced the concept of (Choquet) capacity and the Choquet integral [51]. Capacities can be viewed as nonadditive measures that satisfy some extra assumptions that allow the introduction of the integral with respect to such measures. Choquet capacities and integrals are nonlinear, but they still have some continuity properties. Choquet theory was developed further in a series of papers (see, e.g., [68, 209]).

The latter concept is widely used in decision theory [68, 209] and robust Bayesian inference [120] and is closely related to other concepts modelling different kinds of probability set, like: lower probabilities [88], belief functions [68], lower envelopes [120], lower expectations and lower previsions [227]. A *Choquet capacity* is, roughly speaking, a nonlinear extension of a measure. Its main advantage is that the analytic and numerical theories of capacities are quite well developed. For every space $\mathbf{X}$ and algebra $\mathscr{A}$ of subsets of $\mathbf{X}$, a set function

$$c : \mathscr{A} \to [0, 1]$$

is called a *normalised capacity* if it satisfies the following axioms:

(i) normalisation:

$$c(\emptyset) = 0, \qquad c(\mathbf{X}) = 1;$$

(ii) monotonicity

$$A \subset B \Rightarrow c(A) \leq c(B)$$

for all $A, B \in \mathscr{A}$.

A capacity is called *convex* (or *supermodular*) if in addition to (i) and (ii) it satisfies the property

(iii) for all $A, B \in \mathscr{A}$,

$$c(A \cup B) \geq c(A) + c(B) - c(A \cap B).$$

A special type of convex capacity is the *belief function*, or *lower probability*, presented and discussed by Dempster [68] and Shafer [211].

A capacity is called a *probability* if (iii) holds everywhere with equality, i.e., it is additive. If a capacity satisfies the inverse inequality in (iii), then it is called *submodular* or *strongly subadditive*.

If $\mathbf{X}$ is a topological space, c is a *Choquet capacity* if it is continuous from below, i.e.,

$$c(F_n) \downarrow c(F)$$

for each sequence of closed sets $F_n \downarrow F$.

The Choquet *capacity* associated to a Markov process $\mathbf{M}$ (see [41] and the references therein) is defined as a set function $\mathrm{Cap}_{\mathbf{M}} : (\mathbf{X}, \mathscr{B}) \rightarrow [0, 1]$ such that

$$\mathrm{Cap}_{\mathbf{M}}(B) := \mathbb{P}(T_B < \infty) = \mathbb{P}(T_B < \zeta) \tag{5.11}$$

for all $B \in \mathscr{B}$, where T_B is the first hitting time of B, i.e.,

$$T_B = \inf\{t > 0 | x_t \in B\}.$$

The capacity of a measurable set B can be thought of as a 'measure' of all process trajectories that ever visit B over an infinite horizon time. It is clear that the capacity (with respect to $\mathbf{M}$) of a set B is nothing but the reach probability in infinite horizon time corresponding to B.

According to [89], it can be shown that $\mathrm{Cap}_{\mathbf{M}}$ is

- monotone increasing, i.e.,

$$A \subset B \Rightarrow \mathrm{Cap}_{\mathbf{M}}(A) \leq \mathrm{Cap}_{\mathbf{M}}(B);$$

- submodular, i.e.,

$$\mathrm{Cap}_{\mathbf{M}}(A \cup B) + \mathrm{Cap}_{\mathbf{M}}(A \cap B) \leq \mathrm{Cap}_{\mathbf{M}}(A) + \mathrm{Cap}_{\mathbf{M}}(B);$$

- countably subadditive, i.e.,

$$\mathrm{Cap}_{\mathbf{M}}\left(\sum_{n=1}^{\infty} A_n\right) \leq \sum_{n=1}^{\infty} \mathrm{Cap}_{\mathbf{M}}(A_n).$$

The initial definition (see [89] and the references therein) of this notion describes the capacity $\mathrm{Cap}_{\mathbf{M}}$ as an upper envelope of a non-empty class of probability measures on $\mathscr{B}$. This means that $\mathrm{Cap}_{\mathbf{M}}$ is an *upper probability*.

The connection between the reach probabilities and capacities can be deployed in several ways. One way is to use the characterisation of the capacity associated to a Markov process as solution of an obstacle problem. Then a natural step is to leverage the analytic methods (variational inequalities) available for the obstacle problem in order to characterise the reach probabilities. Another way is to dig deeper in the theory of imprecise probabilities and Choquet capacities to find out computational methods. It seems that a natural approach is to use Bayesian statistics for capacities. We will detail these approaches in separate chapters.

5.5.6   Réduite and Balayage

Reach set probabilities represent a solution for an appropriate Dirichlet boundary value problem. One way to study the existence and uniqueness of a Dirichlet problem is via the balayage method. The *balayage method* is due to Poincaré, who, in 1890, gave the first complete proof, in rather general domains, of the existence and uniqueness of a solution of the Laplace equation for any continuous Dirichlet boundary condition. This is an iterative method that relies on solving the Dirichlet problem on balls in the domain and makes extensive use of the maximum principle and Harnack's inequality for harmonic functions. The word 'balayage' means sweeping or clearing dust away in French. In the classical balayage method, one cleans a domain completely from any mass sitting in it. In the context of the problems treated in this book, the balayage method is characterised based on the leverage of the Perron perspective [192]. Perron proved that for the Laplace operator, the balayage of a measure μ on a given domain D is nothing but the infimum of all superharmonic functions on D having prescribed values on ∂D (usually equal to the Newtonian potential generated by μ).

For Markov processes, the notion of *balayage of excessive functions* is due to Hunt [128–130]. In the following, we briefly summarise the connections between hitting operators, balayage and réduite. Suppose that $\mathscr{E}_\mathbf{M}$ is the cone of excessive functions and $\mathscr{V}$ is the resolvent associated to $\mathbf{M}$. For any $f : \mathbf{X} \to \mathbb{R}_+$, we denote by $\mathbf{R}f$ the function

$$\mathbf{R}f := \inf\{u \in \mathscr{E}_\mathbf{M} | u \geq f\}, \qquad (5.12)$$

called *the réduite of* f, or reduced function associated to f (with respect to the resolvent $\mathscr{V}$). The réduite of f differs from f alone on a negligible set. In potential theory, in the proof of the existence of the réduite for a function f, it is usually assumed that f is the difference of two excessive functions. In Markov process theory, the existence of the réduite of a bounded measurable function $f : \mathbf{X} \to \mathbb{R}$ (with respect to the resolvent $\mathscr{V}$) is proved using ideas different from those used in potential theory. In fact, this existence is based on the following equality [79]:

$$\mathbf{R}f(x) = \sup\left\{\mathbb{E}_x\left[f(x_S)1_{\{S<\infty\}}\right]; S \text{ stopping time}\right\}. \qquad (5.13)$$

For any subset A of $\mathbf{X}$ and $v \in \mathscr{E}_\mathbf{M}$, the function

$$\mathbf{R}_A v = \mathbf{R}(1_A v)$$

is called *the réduite of v on A*, where the operator R is defined by (5.12). We use the convention $0 \cdot (+\infty) = (+\infty) \cdot 0 = 0$.

It is known that for any $A \in \mathscr{B}(\mathbf{X})$ and $v \in \mathscr{E}_\mathbf{M}$ the function $R_A v$ is $\mathscr{B}(\mathbf{X})$-measurable and it is $\mathscr{V}$-supermedian, i.e.,

$$e^{-\alpha t} P_t[\mathbf{R}_A v] \leq \mathbf{R}_A v$$

for all $t \geq 0$ and $\alpha > 0$. In this case, we denote by $B_A v$ the $\mathscr{V}$-excessive regularisation of $R_A v$, i.e.,

$$B_A v := \sup_\alpha \alpha V_\alpha[\mathbf{R}_A v].$$

$B_A v$ is called also *balayage of the excessive function v on A*.

Moreover, $B_A v$ is the *lower semicontinuous regularisation* of $R_A v$ with respect to a suitable topology on X (namely, the *fine topology*, which is the smallest topology that makes all excessive functions continuous). For relations between v, $R_A v$ and $B_A v$, one can consult, for example, [108]. Briefly, these relations are:

- on the whole state space X, we have

$$v \geq \mathbf{R}_A v \geq B_A v$$

- on the complementary set of A, we have

$$\mathbf{R}_A v = B_A v \quad \text{on } \mathbf{X} \backslash A \tag{5.14}$$

- moreover, the equality

$$v = \mathbf{R}_A v = B_A v \tag{5.15}$$

holds also on $A \backslash N$, where N is a negligible[1] set of A.

A way to reformulate the reduction idea is to say that $B_A v$ solves an *obstacle problem*: $B_A v$ is the smallest excessive function above the obstacle described by $1_A v$.

In [128–130], it has been proved that the balayage of an excessive function v on a Borel set A is nothing but the *hitting operator* given by (5.6) of the underlying Markov process. Formally, this can be summarised' as follows.

Proposition 5.8 *The balayage of an excessive function v on a Borel set A is given by $P_A v$, where P_A is the hitting operator associated to the underlying Markov process (x_t), i.e.,*

$$B_A v = P_A v = \mathbb{E}_x\{v \circ x_{T_A} | T_A < \infty\},$$

where T_A is the first hitting time of A given by (5.4).

Since for a right Markov process, the function identically equal to 1 is excessive, the following characterisation of the stochastic reachability is straightforward.

---

[1] In fact, N is the set of irregular points of A, which is a semipolar set, see [108].

Proposition 5.9 *For any* $x \in \mathbf{X}$ *and Borel set* $A \in \mathscr{B}(\mathbf{X})$, *we have*

$$\mathbb{P}_x\big[Reach_\infty(A)\big] = B_A 1(x) = P_A 1(x) = \mathbb{P}_x[T_A < \zeta].$$

At this stage, we have the connection between

- stochastic reachability, which is a probabilistic concept defined in terms of hitting distributions, and
- the balayage/réduite, which is a potential theory concept defined using the semi-group/resolvent of operators associated to the given Markov process.

This connection plays a prominent role in the unearthing of analytic solutions for the estimations of the reach set probabilities. The réduite concept is related to optimal stopping problems for stochastic processes. Then, variational and quasi-variational inequalities can be derived. As we have seen, the balayage concept is also linked with the obstacle problem. Therefore, we can prove further that the problems of finding the reach probabilities are equivalent to physical problems of minimising an energy integral.

5.6 Some Remarks

In this chapter, we have set up the theoretical foundations of the analysis framework for stochastic reachability. Due to different characterisations of this concept, the estimation of the reach set probabilities can be tackled via three main avenues:

- the purely probabilistic approach, which is built on the basis of hitting/exit distributions, martingale characterisation, resolvent/semigroup of operators;
- the analytic approach, which starts with the connections between reachability and potential theory (réduite, balayage) and the characterisation of Markov processes using Dirichlet forms;
- the statistical approach, which considers imprecise probabilities involvement when one defines stochastic reachability.

These approaches are discussed in the following three chapters. Then, further developments follow in the last three chapters.

Chapter 6
Probabilistic Methods for Stochastic Reachability

6.1 Overview

This chapter is dedicated to the purely probabilistic treatment of stochastic reachability. The backbone of this approach is built on traditional and less traditional methods used in the study of Markov processes like transition probabilities, resolvent, infinitesimal generator, martingale characterisation, potential theory concepts (Dirichlet forms, capacity) and so on. Usually, these methods are more comprehensible for the case of simpler Markov processes (like Markov chains, Wiener or diffusion processes). Therefore, whenever possible, we exemplify the proposed approach for these simpler processes. The idea is that stochastic hybrid systems, at different levels of abstraction/approximation, can be related also with these simpler processes. The hybrid behaviour of such systems has proved always to be the main complexity aspect when we are studying reachability problems. Then, our expectations regarding these problems is to obtain accurate approximation results or lower/upper bounds of the reach probabilities.

6.2 Reachability Estimation via Hitting Times

In this section, reachability questions are characterised as hitting probability distributions. Before we treat the general case of stochastic hybrid systems, we will discuss first some particular cases of continuous/discrete time Markov chains and diffusion processes. The reason for this is the fact that for Markov chains and diffusions there exist quite well-developed methods and algorithms for stochastic reachability described as a first passage problem. Later, we will discuss how and when these methods can be extended to stochastic hybrid systems.

6.2.1 Hitting Distributions for Markov Chains

Note that the first hitting times are known in the literature as *first passage times*. Then, the computation of reach set probabilities can be reduced to the computation

L.M. Bujorianu, *Stochastic Reachability Analysis of Hybrid Systems*,
Communications and Control Engineering,
DOI 10.1007/978-1-4471-2795-6_6, © Springer-Verlag London Limited 2012

of some probability distributions of first passage times, known also as the first passage problem. This is not an easy problem, even for simpler models like Markov chains. In the following, we briefly summarise the 'classical' methods existing for the computations of the first passage time distributions for simpler Markov models. The interested reader can consult [219] for a complete exposition of passage times for Markov chains.

First Passage Time Distributions for DTMCs   Let $\{X_n : n \geq 0\}$ be a DTMC with state space $\mathbb{S} = \{0, 1, 2, \ldots\}$, the transition probability matrix P and the initial distribution p. Denote

$$T = \inf\{n \geq 0 | X_n = 0\}. \tag{6.1}$$

The random variable T is called the *first passage time* into state 0 (or the *first hitting time* of $\{0\}$).

We can associate to T two probability distributions:

$$\begin{aligned} \alpha_i(n) &= \mathbb{P}(T = n | X_0 = i), \\ u_i(n) &= \mathbb{P}(T \leq n | X_0 = i). \end{aligned} \tag{6.2}$$

The following theorem provides a recursive method of computing $u_i(n)$.

Theorem 6.1

$$u_i(n) = p_{i0} + \sum_{j=1}^{\infty} p_{ij} u_j(n-1) \quad \text{for all } i, n \geq 1$$

with

$$u_i(0) = 0 \quad \text{for all } i \geq 1.$$

Now we will study methods of computing the so-called probability of visiting state 0:

$$u = \mathbb{P}\{X_n = 0 \text{ for some } n \geq 0\} = \mathbb{P}\{T < \infty\}$$

where T is the first passage time of 0, given by (6.1).

In many applications, the state 0 is an absorbing state, i.e., once X_n visits 0 it cannot leave it. The quantity u is called the *absorption probability*. We may define this absorption probability with respect to an initial state i as

$$u_i = \mathbb{P}\{T < \infty | X_0 = i\}, \quad i \geq 1.$$

Then

$$u_i = \lim_{n \to \infty} u_i(n),$$

where $u_i(n)$ is given by (6.2).

Theorem 6.2 *The $\{u_i : i \geq 1\}$ is the smallest solution of the equation*

$$u_i = p_{i0} + \sum_{j=1}^{\infty} p_{ij} u_j \quad for \ all \ i \geq 1.$$

First Passage Time Distributions for CTMCs   Consider a finite stationary time-homogeneous irreducible CTMC with n states $\{1, 2, \ldots, n\}$ and $n \times n$ generator matrix Q. The chain is called *irreducible* if for any pair i, j of states we have $p_{ij}(t) > 0$ for some t. If $X(t)$ denotes the state of the CTMC at time $t \geq 0$, then the first passage time from a source state i into a different state j is

$$T_j = \inf\{u > 0 : X(u) = j | X(0) = i\}.$$

Let S be the first time when the chain leaves the state i. Recall that S is exponentially distributed, i.e.,

$$S \sim \exp(-q_{ii}).$$

Then by the Markov property we have

$$\mathbb{E}(T_j | X(0) = i) = \mathbb{E}(S | X(0) = i) + \sum_{k \neq i, j} \mathbb{E}(T_j | X(0) = k) \mathbb{P}(X(S) = k | X(0) = k).$$

The sum does not include i because the chain cannot leave i to arrive also in i and it does not include j because $\mathbb{E}(T_j | X(0) = j) = 0$. We know that

$$\mathbb{E}(S | X(0) = i) = \frac{1}{-q_{ii}},$$

$$\mathbb{P}(X(S) = k | X(0) = k) = \frac{q_{ik}}{(-q_{ii})}, \quad k \neq i,$$

and then

$$\mathbb{E}(T_j | X(0) = i) = (-q_{ii})^{-1} \left\{ 1 + \sum_{k \neq i, j} q_{ik} \mathbb{E}(T_j | X(0) = k) \right\}.$$

Using the notation

$$u_i := \mathbb{E}(T_j | X(0) = i),$$

we obtain

$$u_i(-q_{ii}) = 1 + \sum_{k \neq i, j} q_{ik} u_k$$

or

$$1 + \sum_{k \neq, j} q_{ik} u_k = 0.$$

Let us denote the matrix obtained from Q by deleting the row and column corresponding to state j by $Q(j)$. Then for all possible starting values $i \neq j$, the equation above can be written in matrix form as

$$1 + Q(j)\mathbf{u} = \mathbf{0}, \tag{6.3}$$

where $\mathbf{1} = (1, 1, \ldots, 1)^\mathsf{T}$ and $\mathbf{u} = (u_1, u_2, \ldots, u_n)^\mathsf{T}$. Then the solution for (6.3) is

$$\mathbf{u} = \left[-Q(j)\right]^{-1}\mathbf{1}.$$

For CTMC, well studied methods for the computation of the first passage time distributions exist in both computer science and applied probabilities. The most popular methods are the *Laplace transform method* and the *uniformisation method*.

Let us consider a different method for the computation of the densities corresponding to the first passage times based on uniformisation. Suppose now that J is a nonempty set of targets and its corresponding first passage time is

$$T_{iJ} = \inf\{u > 0 : X(u) \in J \,|\, X(0) = i\}.$$

T_{iJ} is a random variable with an associated probability density function denoted by $f_{iJ}(t)$. To determine $f_{iJ}(t)$, we need to convolve the exponentially distributed sojourn times over all possible paths including cycles from i to the set J.

Suppose that we have constructed the uniformised DTMC $\{\widehat{X}_n\}_n$ with the stochastic transition matrix $\widehat{P}$ given by (2.6). Let us add an extra absorbing state δ to the uniformised chain, which is the sole successor state for all target states. In this chain, the time needed for n transitions along a path will have the Erlang distribution with parameters n and q. Denote by $\widetilde{P}$ the one-step transition matrix of the modified uniformised chain. Then

$$f_{iJ}(t) = \sum_{n=1}^{\infty} \frac{q^n t^{n-1} e^{-qt}}{(n-1)!} \sum_{k \in J} \pi_k^{(n)},$$

where

$$\pi^{(n+1)} = \pi^{(n)} \widetilde{P} \quad \text{for } n \geq 0$$

with

$$\pi_k^{(0)} = \begin{cases} 0, & \text{for } k \neq i \\ 1, & \text{for } k = i. \end{cases}$$

The corresponding passage time cumulative distribution function is given by

$$F_{iJ}(t) = \sum_{n=1}^{\infty} \left[\left(1 - e^{-qt} \sum_{k=0}^{n-1} \frac{(qt)^k}{k!}\right) \sum_{k \in J} \pi_k^{(n)} \right].$$

First Passage Time for Diffusions   The first passage time (FPT) problem has seen more than a century of history, starting with Bachelier [10], who examined the first passage of the Wiener process to the constant boundary. His work was expanded by Paul Lévy to general linear boundary. For general diffusions, FPT problems were investigated for the first time in the work of Khinchine [141], Kolmogorov [144] and Petrowsky [191]. Foundations of the general theory of Markov processes were set up by Kolmogorov [144] in 1931. His work clarified the deep connection between probability theory and mathematical analysis and initiated the partial differential equations approach to the FPT problem.

For diffusion processes, the main tools for dealing with the first passage time problems are:

- partial differential equations (PDEs),
- space and time change,
- measure change and the martingale approach via the optional sampling theorem.

For the FPT problem, formulation in the PDE setting is done through the (Fokker–Planck) Kolmogorov forward equation.

If the process is a continuous diffusion with infinitesimal generator L, and the target set A is a closed set, it is known that if the PDE (Dirichlet problem)

$$\begin{cases} \dfrac{\partial u}{\partial t} = Lu & \text{on } E \times (0, T] \\ u = 0 & \text{on } E \times \{0\} \\ u = \mathbf{1}_{\partial E} & \text{on } \partial E \times (0, T] \end{cases}$$

has a bounded solution, then

$$u(x, t) = \mathbb{P}_x\{T_A \leq t, x_{T_A} \in \partial E\}, \quad 0 \leq t \leq T. \tag{6.4}$$

Here, E is the complementary set of A.

The space/time change approach is based on the reduction of processes of the form

$$x_t = g(W_{B(t)}; t)$$

into a Wiener process after the inversion of g and B. Examples of such processes include the Ornstein–Uhlenbeck process, Brownian bridge and geometric Brownian motion. Continuous Gauss–Markov processes also represent a subclass of space/time-changed Brownian motions, which was used in [75] to obtain an explicit relationship between the FPT distribution of x_t and the corresponding one associated to a Wiener process. Such processes belong to a bigger class of diffusion processes that can be transformed into Wiener processes. Such a transformation can be realised by reducing the Kolmogorov equation corresponding to a diffusion process to the backward equation for the Wiener process, according to [199]. This reference provides alternative conditions under which a diffusion process can be transformed into a Wiener process. Such transformations were generalised in [50], where an equivalence concept between diffusion processes is introduced.

Using appropriate martingales, one can use a measure change method (rescale the underlying probability measure) to transform the given diffusion into a martingale. Then the optional sampling theorem represents a versatile tool for obtaining integral equations from the FPT problems.

6.2.2 Hitting Distributions of GSHSs

In this subsection, we extend the approach developed in [42] for piecewise deterministic Markov processes, for more general stochastic hybrid processes. This is possible since a GSHS represents a 'diffusion version' of a PDMP and we have seen that the main properties of a GSHS and the expression of its infinitesimal generator are quite similar to those corresponding to a PDMP.

Let

$$H = \big((Q, d, \mathscr{X}), b, \sigma, \text{Init}, \lambda, R\big)$$

be a GSHS with the hybrid state space $\mathbf{X}$. Let us consider a fixed starting point $x_0 \in \mathbf{X}$ and the sequence

$$T_1 < T_2 < \cdots < T_k < \cdots$$

of jump times associated with x_0. Let A be a Borel subset of $\mathbf{X}$, thought of as a target set in the context of the reachability problem. Let

$$T = T_A$$

be the first hitting time of A.

The following lemma is extremely useful for estimating the expectations of the moments of hitting times.

Lemma 6.3 [77] *Suppose that*

$$m(t) := \sup\{\mathbb{P}_x[T > t], x \in \mathbf{X}\}.$$

Then, for all $t \geq 0$, $x \in \mathbf{X}$ we have

$$\mathbb{E}_x T \leq \frac{t}{1 - m(t)}.$$

Moreover, if $m < 1$, for

$$0 < p < -\frac{1}{t} \ln m$$

$\mathbb{E}_x e^{pT}$ *is an analytic function of p in a neighbourhood of the origin, i.e., we have*

$$\mathbb{E}_x e^{pT} = \sum_{k=1}^{\infty} \frac{p^k}{k!} \mathbb{E}_x T^k.$$

Proposition 6.4 *Suppose that T_A is finite and there exists T_k such that*

$$T_A \in [T_k, T_{k+1}).$$

Then the expectation of the first hitting time of the target set A can be estimated as follows

$$\mathbb{E}_{x_0} T_A \leq \frac{T_{k+1}}{1 - \mathbb{P}_{x_{T_k}}[T_A > T_{k+1}]} \leq \frac{T_{k+1}}{\exp[-\int_{T_k}^{T_A} \lambda(\phi(x_{T_k}, \tau))\, d\tau]}.$$

Proof The proof is straightforward and is based on Lemma 6.3 and the definition of the survivor function, which has been used in defining the execution of a GSHS. □

Practically, if we are able to compute the quantities

$$\tau_k := \frac{T_k}{\mathbb{E}_{x_0} T_A},$$

$k = 0, 1, 2, 3, \ldots$ (with the convention $T_0 = 0$), then

$$T_A \in [T_{k_0}, T_{k_0+1}),$$

where τ_{k_0} is the biggest one in the family $\{\tau_k\}$. This gives information regarding the number of steps k_0 after the set A is reached.

This method provides information only on the first order moments of the hitting time T_A. Assuming that for a stochastic hybrid process, one would have information regarding the jumping times, then recurrent algorithms could be constructed to estimate the reach set probabilities. This requires, of course, more information about the Markov chain embedded in the structure of a stochastic hybrid system. Note that for PDMPs, the properties of such a Markov chain are well studied [64, 65]. For a general stochastic hybrid system, the embedded Markov chain is more complex and its 'states' are diffusion paths of its components. Then the analysis of such a chain is not much easier than the analysis of the stochastic hybrid process itself. Research along this avenue can produce more if we consider a slightly different model for stochastic hybrid systems. Such a model is built on the skeleton of a Markov decision process and the continuous evolutions in modes represent the 'actions' of the decision process. In this way, the reachability problem should have two levels of investigation:

- one at the bottom level: studying the continuous trajectories in modes (hitting probabilities for continuous processes); and
- one at the decision level: studying the discrete transitions (first passage time distributions for discrete processes).

6.3 Reachability Estimation via Transition Semigroup

In this section, we start with the techniques available for Markov chains for computing the réduite/value function that represents the desired reachability measure. These techniques use the one-step probabilities (or the stochastic matrix) that define a Markov chain. Then we extend these techniques to continuous-time continuous-space Markov processes, in particular to stochastic hybrid systems.

In the following, we present some algorithms used in probabilistic potential theory for the computation of the réduite. We start with the simple case of Markov chain, then the technique is extended to more complex Markov processes.

6.3.1 The Case of a Markov Chain

Let $\{x_n | n = 0, 1, 2, \ldots\}$ be a discrete-parameter Markov chain defined on a probability space $(\Omega, \mathscr{F}, \mathbb{P})$ with values in $\mathbf{X}$ and let its natural filtration be

$$\mathscr{F}_n = \sigma(x_k | k \leq n).$$

A stochastic kernel $k(x, \Gamma)$ is a *transition function for a time-homogeneous Markov chain* (x_n) if

$$\mathbb{P}(x_{n+1} \in \Gamma | \mathscr{F}_n) = k(x_n, \Gamma).$$

The *potential kernel* associated to k (or to the Markov chain) is defined by

$$U := \sum_{n=0}^{\infty} k^n \quad (k^0 = I). \tag{6.5}$$

For Markov chains, the definition of the excessive function is simplified. A measurable positive function f is called *excessive* with respect to k if

$$f \geq kf,$$

where

$$kf(x) := \int f(y)k(x, dy).$$

The *reduced function*, or *réduite*, of a bounded measurable function f is defined by (5.12).

Proposition 6.5 (Reduite Algorithm) [35] *Let g be a measurable and bounded real-valued function on* $\mathbf{X}$. *Let* $\{g_n\}$ *be the sequence defined recurrently as*

$$g_0 = g^+,$$

$$g_{n+1} = g_n \vee k u_n.$$

Then the sequence $\{g_n\}$ *is increasing and converges to* $\mathbf{R}g$, *i.e.*,

$$g_n \nearrow \mathbf{R}g.$$

It can be proved that $\mathbf{R}g$ is nonlinear, measurable and

$$\mathbf{R}g = g \vee \mathbf{R}g$$

and, if

$$x \in [g < \mathbf{R}g] \cup [g \leq 0],$$

then

$$\mathbf{R}g(x) = k(\mathbf{R}g)(x).$$

Moreover, the *réduite operator* $\mathbf{R}$ is

- increasing:

$$g \leq v \Rightarrow \mathbf{R}g \leq \mathbf{R}v$$

- subadditive:

$$\mathbf{R}(g + v) \leq \mathbf{R}g + \mathbf{R}v$$

- positively homogeneous:

$$\mathbf{R}(tu) = t\mathbf{R}g, \quad t \geq 0$$

- continuous on ascending sequences:

$$g_n \nearrow g \Rightarrow \mathbf{R}g_n \nearrow \mathbf{R}g.$$

6.3.2 The Case of a Markov Process

Let us consider a Markov process

$$\mathbf{M} = (x_t, \mathbb{P}_x)$$

with all the operator semigroup $\mathscr{P} = (\mathbf{P}_t)_{t>0}$ and with excessive function cone $\mathscr{E}_{\mathbf{M}}$. For $t > 0$ and g a positive measurable bounded function, we write $\mathbf{R}_{(t)}g$ for the réduite of g with respect to the kernel $\mathbf{P}_t$.

Using the extended generator $\mathscr{L}$ of the semigroup $\mathscr{P}$, we may approximate the réduite of a function g that belongs to the domain of the extended generator $D_e(\mathscr{L})$.

Note that there exists an equivalent definition that makes use of the operator resolvent defined by (2.25). A function $g \in \mathscr{B}^b(\mathbf{X})$ is said to belong to the domain of the extended generator of the semigroup $\mathscr{P}$ if there exists another Borel measurable function h such that

$$V_\alpha|h| < \infty \quad \text{and} \quad g = V_\alpha(\alpha g - h), \quad \alpha > 0.$$

The function h is unique up to changes on a set of potential zero ($A \in \mathscr{B}$ is of potential zero if $V_\alpha 1_A = 0, \alpha > 0$). Explicitly, we have

$$h = \lim_{n \to \infty} n(n V_n g - g).$$

Proposition 6.6 [35] *Let* $g \in D_e(\mathscr{L})$ *with* $\mathscr{L}g$ *bounded and* $t \geq 0$. *Then*

(i) $\mathbf{R}g = \lim_{n \to \infty} \uparrow \mathbf{R}_{(t/2^n)}g$

(ii) $\mathbf{R}g - \mathbf{R}_{(t)}g \leq t\|\mathscr{L}g\|_\infty.$

The last result can be very useful when the expression of the operator semigroup is known, as in the case of diffusion processes.

6.3.3 The Case of a Stochastic Hybrid Process

Now we return to the framework of GSHS. The executions of such systems are complex Markov models that allow forced (predictable) jumps. The existence of these predictable jumps involves a peculiarity of the extended generator. This is reflected in the boundary condition (4.22), which should be satisfied by all functions belonging to the domain of this generator. It is clear that the indicator functions of some measurable sets do not belong to this domain. Therefore, it is clear that the approximation of the réduite corresponding to such indicator functions cannot follow the scheme proposed in the Proposition 6.6. Then the natural idea is to approximate the whole process by Markov chains and to use convergence results available for the réduite. For stochastic hybrid processes, we will use approximations with Poisson time stepping rather than approximations with fixed time stepping (Euler–Maruyama). The reason is that the discrete jumps can be captured in a better way when the sampling time is given in a Poisson fashion.

Let $\mathbf{M} = (x_t, \mathbb{P}_x)$ the Markov process describing the behaviour of a SHS, H. The construction of an approximation sequence of jump processes for $\mathbf{M}$ needs

1. A sequence of homogeneous Markov chains (α^n). Each $\alpha^n = (\alpha_k^n)_{k=0,1,2,...}$ is a Markov chain with some initial distribution v and the transition probability function K_n, defined by

$$K_n(x, dy) := n V_n(x, dy) \qquad (6.6)$$

   where V_n is the stochastic kernel computed from formula (2.25).
2. A sequence of Poisson processes (θ^n). Each $\theta^n = (\theta_t^n)_{t \geq 0}$ is a Poisson process[1] with the parameter n, independent of α^n.

Using these ingredients, we then define, for each $n \geq 1$, a *continuous-time Markov jump process* as follows:

$$\rho_t^n := \alpha_{\theta_t^n}^n, \quad t \geq 0. \qquad (6.7)$$

This means that the jump times of the process (ρ_t^n) are given by the arrival times of the Poisson process (θ_t^n), and its values between jumps are provided by the Markov chain (α_k^n). The infinitesimal generator of (ρ_t^n) can be expressed as follows [83]:

$$L^n u(x) = n \int_{\mathbf{X}_\Delta} \big[u(y) - u(x) \big] n V_n(x, dy)$$

where $\mathbf{X}_\Delta = \mathbf{X} \cup \{\Delta\}$ (where Δ is thought of as a cemetery/deadlock point of the process $\mathbf{M}$). It can be shown that (ρ_t^n) converges to (x_t) with respect to the Skorokhod topology (defined on the space of càdlàg functions).

The following convergence result for the reach set probabilities does not depend on the characterisation of these measures as appropriate réduite/value functions.

---

[1] I.e., $P(\theta_t^n = k) = \exp(-nt) \frac{(nt)^k}{k!}$.

Theorem 6.7 [47] *If the sequence $\{\rho_t^n\}_{n\geq 1}$ of strong Markov processes converges weakly to (x_t) (under $\mathbb{P}_x$) as $n \to \infty$, then for any measurable set A*

$$\mathbb{P}_x^n\big[Reach_\infty^n(A)\big] \to \mathbb{P}_x\big[Reach_\infty(A)\big].$$

The main result of this section is the following:

Theorem 6.8 *For a measurable set A, the reach set probabilities can be approximated as follows:*

$$\mathbb{P}_x\big[Reach_\infty(A)\big] = \lim_{n\to\infty} \mathbf{R}_{(n)} 1_A$$

where $\mathbf{R}_{(n)}$ is the réduite operator corresponding to the jump kernel K_n given by (6.6).

Using Proposition 6.5 (réduite algorithm), the réduite $\mathbf{R}_{(n)} 1_A$ corresponding to the stochastic kernel K_n can be computed as follows:

$$\mathbf{R}_{(n)} 1_A = \lim_{m\to\infty} \uparrow g_m$$

where

$$g_0 := 1_A,$$
$$g_{m+1} := g_m \vee K_n g_m.$$

We remark that, in using this algorithm, it is essential we know the expression of the stochastic kernel K_n. It can be proved that

$$K_n := n V_n = n(nI - \mathscr{L})^{-1}, \quad n \geq 1,$$

where I is the identity operator [83]. Moreover, V_n is the *potential kernel* of the process $\mathbf{M}$ killed with the exponential rate n. Therefore, further approximations of the generator $\mathscr{L}$ can be employed to obtain evaluations of the reach set probabilities with different degrees of accuracy.

6.4 Reachability Estimation via Martin Capacities

In this section, we estimate reach set probabilities and by a capacity function with respect to a scale-invariant modification of the kernel operator. We start by defining the Martin capacity, which is an imprecise probability measure defined with respect to an energy form. Then we present some classical results regarding estimation of the hitting probabilities for Markov chains and Brownian motion, based on the Martin capacity. This capacity is defined with respect to an energy form derived from the kernel operator. Finally, we show how these results can be extended to more general Markov processes and eventually to stochastic hybrid processes.

6.4.1 Martin Capacity

Let Λ be a set and $\mathscr{B}$ a σ-algebra of subsets of Λ. Given a measurable function

$$F : \Lambda \times \Lambda \to [0, \infty]$$

and a finite measure μ on $(\Lambda, \mathscr{B})$, the F-energy of μ is

$$F(\mu) := F(\mu, \mu) = \int_\Lambda \int_\Lambda F(\alpha, \beta)\, d\mu(\alpha)\, d\mu(\beta).$$

The Martin *capacity with respect to F* is

$$\mathrm{Cap}_F(\Lambda) = \left[\inf F(\mu)\right]^{-1}, \tag{6.8}$$

where the infimum is over probability measures μ on $(\Lambda, \mathscr{B})$ and by convention, $\infty^{-1} = 0$.

If Λ is included in a Euclidean space, then $\mathscr{B}$ can be taken as the Borel σ-algebra. If Λ is countable, then $\mathscr{B}$ will be the σ-algebra of all its subsets.

6.4.2 The Case of a Markov Chain

Let (X_n) be a DTMC with the state space $\mathbf{S} = \{0, 1, 2, 3, \ldots\}$ and $P = (p_{ij})_{i,j \in \mathbf{S}}$ its *one-step transition matrix*. Then its kernel operator (or Green operator) is

$$U(i, j) = \sum_{n=0}^\infty p_{ij}^{(n)} = \sum_{n=0}^\infty \mathbb{P}_i[X_n = j],$$

where $p_{ij}^{(n)}$ are the n-step transition probabilities and $\mathbb{P}_i$ is the law of the chain when the initial state is i.

We want to estimate the reach probability of a target set Λ for the given DTMC. Assume that the Markov chain is transient, i.e.,

$$U(i, j) < \infty,$$

for $i, j \in \mathbf{S}$.

Proposition 6.9 [20] *Let (X_n) be a DTMC with state space $\mathbf{S}$ having initial state i_0 and transition probabilities (p_{ij}). For any subset Λ of $\mathbf{S}$, we have*

$$\frac{1}{2}\mathrm{Cap}_K(\Lambda) \le \mathbb{P}_{i_0}\left[Reach_\infty(\Lambda)\right] \le \mathrm{Cap}_K(\Lambda),$$

where K is the Martin kernel (with respect to the initial state i_0) defined by

$$K(i, j) := \frac{U(i, j)}{U(i_0, j)}. \tag{6.9}$$

6.4.3 The Case of Brownian Motion

Results from the previous subsection can be easily extended to the case of the Wiener process (Brownian motion) (W_t) defined on the Euclidean space $\mathbb{R}^d$, $d \geq 3$. In this case, the Green operator is given as

$$U(x, y) = \|x - y\|^{2-d},$$

and the Martin kernel is

$$K(x, y) := \frac{\|y\|^{d-2}}{\|x - y\|^{d-2}} \tag{6.10}$$

for $x \neq y$ and $K(x; x) = \infty$.

If $\mathbb{P}_x(\cdot)$ denotes the probability (Wiener measure) when all paths issued from the point x for the standard Wiener process in $\mathbb{R}^3$; Λ is a compact set (the conductor body); $T_\Lambda(\omega)$ is the hitting time of Λ by the path ω, then a classical result says that

$$\mathbb{P}_x(T_\Lambda < \infty) = \int_{\partial \Lambda} g(x, \beta) \mu_\Lambda(d\beta), \tag{6.11}$$

where $\partial \Lambda$ is the boundary of Λ; $g(x, \beta)$ is the associated *potential density*

$$g(x, \beta) = \frac{1}{2\pi \|x - \beta\|},$$

and μ_Λ is called an *equilibrium measure*. On the other hand, the probability that a Brownian motion will ever visit a given set Λ, which appears on the left hand side of (6.11), can be classically estimated using the capacity of Λ with respect to the Green kernel $g(x, \beta)$. In [20], it is shown that replacing the Green kernel by the Martin kernel $g(x, \beta)/g(0, \beta)$ yields improved estimates, which are exact up a factor of 2.

Proposition 6.10 [20] *Let Λ be any closed set in $\mathbb{R}^d$, $d \geq 3$. Then the reach set probability corresponding to (W_t) and Λ can be estimated as follows:*

$$\frac{1}{2}\mathrm{Cap}_K(\Lambda) \leq \mathbb{P}_0\big[Reach_\infty(\Lambda)\big] \leq \mathrm{Cap}_K(\Lambda), \tag{6.12}$$

where $\mathbb{P}_0$ is the law of Brownian motion under $W_0 = 0$, and K is the kernel defined by (6.10).

The constants $1/2$ and 1 in (6.12) are sharp.

6.4.4 The Case of a Markov Process

Chung in [53] extended formula (6.11) for temporally homogeneous transient Markov processes $\{x_t, t \geq 0\}$, taking values in a topological space $\mathbf{X}$ which is locally compact and has a countable base with its Borel σ-algebra $\mathscr{B}(\mathbf{X})$. The processes also have the càdlàg property. It is natural to put the problem of generalisation of the above Brownian motion result to more general Markov processes, using Chung's result.

Throughout this section

$$\mathbf{M} = (\Omega, \mathscr{F}, \mathscr{F}_t, x_t, \mathbb{P}_x)$$

will be a Borel right Markov process on $(\mathbf{X}, \mathscr{B})$. In addition, we suppose that $\mathbf{M}$ has the càdlàg property and that $\mathbf{M}$ is *transient*. The transience means that there exists a strictly positive Borel function q such that Uq is bounded, where

$$Uf = \int_0^\infty \mathbf{P}_t f \, dt$$

is the *kernel operator* of the operator semigroup $\mathscr{P} = (\mathbf{P}_t)_{t>0}$.

One can take the sample space Ω for $\mathbf{M}$ to be the set of all paths

$$(0, \infty) \ni t \mapsto \omega(t) \in \mathbf{X}_\Delta$$

such that

(i) $t \mapsto \omega(t)$ is $\mathbf{X}$-valued and càdlàg on $(0, \zeta(\omega))$ where

$$\zeta(\omega) := \inf\{s > 0 \,|\, \omega(s) = \Delta\},$$

(ii) $\omega(t) = \Delta$ for all $t \geq \zeta(\omega)$ and
(iii) $\zeta(\omega) < \infty$.

In this way, $\mathbf{M}$ is realised as the coordinate process on Ω:

$$x_t(\omega) = \omega(t), \quad t > 0.$$

We complete the definition of $\mathbf{M}$ by declaring $x_0(\omega) = \lim_{t \searrow 0} \omega(t), t > 0$.

Because of the transience condition, it is possible to construct a probability measure $\mathbb{P}$ on $(\Omega, \mathscr{F}_t^0)$ under which the coordinate process $(x_t)_{t>0}$ is Markovian with the transition semigroup $(\mathbf{P}_t)_{t\geq 0}$ and one-dimensional distributions

$$\mathbb{P}(x_t \in A) = \mu_t(A), \quad \forall A \in \mathscr{B}, t > 0,$$

where $(\mu_t)_{t>0}$ is an appropriate entrance law (see [89] and references therein).

Let

$$p_t(x, B) = \mathbb{P}_x(x_t \in B) = \mathbf{P}_t 1_B(x)$$

be the transition function associated to the given Markov process, where $t > 0$, $x \in \mathbf{X}$ and $B \in \mathscr{B}(\mathbf{X})$.

Assumption 6.1 All the measures $p_t(x, \cdot)$ are *absolutely continuous* with respect to a σ-finite measure μ on $(\mathbf{X}, \mathscr{B}(\mathbf{X}))$.

We denote the Radon–Nycodim derivative of $p_t(x, \cdot)$ by $\rho_t(x, \cdot)$, i.e.,

$$\rho_t(x, y) = \frac{p_t(x, dy)}{\mu(dy)}.$$

This can be chosen to be measurable in x, y and to satisfy

$$\int_{\mathbf{X}} \rho_s(x, y)\mu(dy)\rho_t(y, z) = \rho_{t+s}(x, z).$$

A σ-finite measure μ on $(\mathbf{X}, \mathscr{B}(\mathbf{X}))$ is called a *reference measure* if

$$\mu(B) = 0 \Leftrightarrow p_t(x, B) = 0$$

for all t and x. Throughout this section we suppose that μ, in Assumption 6.1, is a reference measure.

Remark 6.11 [33] If all p-excessive functions ($p > 0$) associated to the underlying semigroup are lower semicontinuous, then there exists a reference measure

$$\mu = \sum_{n=1}^{\infty} \frac{1}{2^n} \varepsilon^{\{y_n\}}$$

where $\{y_n\}$ is a countable dense subset of $\mathbf{X}$ and $\varepsilon^{\{y_n\}}$ are Dirac measures.

We define the *Green measure* g_x, $x \in \mathbf{X}$, as

$$g_x(B) = \int_0^\infty p_t(x, B)\,dt,$$

which is also the expectation $E^x(\eta_B)$ of the *occupancy time* of $B \in \mathscr{B}(\mathbf{X})$, defined

$$\eta_B := \int_0^\infty 1_{\{x_t \in B\}}\,dt.$$

We define the *Green kernel* as

$$u(x, y) := \int_0^\infty \rho_t(x, y)\,dt$$

and the Martin kernel (with respect to an initial state x_0) as

$$K(x, y) = \frac{u(x, y)}{u(x_0, y)}. \tag{6.13}$$

It is clear that, if $(\mathbf{P}_t)$ is the semigroup of operators associated to the given Markov process, then we get

$$\mathbf{P}_t\varphi(x) = \mathbb{E}_x\varphi(x_t) = \int_{\mathbf{X}} \rho_t(x, \beta)\varphi(\beta)\mu(d\beta)$$

and the *kernel operator*

$$U\varphi(x) = \mathbb{E}_x\left[\int_0^\infty \varphi(x_t)\,dt\right] = \int_{\mathbf{X}} u(x, \beta)\varphi(\beta)\mu(d\beta)$$

where φ is any positive Borel measurable function.

In [53] the following assumptions on the Green kernel u are necessary to prove the desired formula:

Assumption 6.2

(i) $y \to u(x, y)^{-1}$ is finite continuous, for $y \in X$;
(ii) $u(x, y) = +\infty$ if and only if $x = y$.

For a target set A we define a random variable $\gamma_A < \infty$ (M is transient), called the *last exit time* from A as follows:

$$\gamma_A(\omega) = \begin{cases} \sup\{t > 0 | x_t(\omega) \in A\} & \text{if } \omega \in Reach_\infty(A) \\ 0 & \text{if } \omega \in \Omega \backslash Reach_\infty(A). \end{cases}$$

The last exit time does not belong to the standard equipment of a Markov process, because it is not a stopping time (see [54] for further comment). It then follows that

$$x_{\gamma_A^-} \in \overline{A} \quad \text{a.s.}$$

Then the *distribution of the last exit position* $x_{\gamma_A^-}$ is given by

$$L^A(x, B) = \mathbb{P}_x(\gamma_A > 0; x_{\gamma_A^-} \in B)$$

for all $x \in X$, $B \in \mathscr{B}(X)$.

Under these assumptions, there exists an equilibrium measure μ_A, which is σ-finite, concentrated in $\overline{A}$ such that

$$\mu_A(dy) = L^A(x, dy)u(x, y)^{-1}, \quad \forall x \in X.$$

For a transient set A we have

$$\{0 \leq T_A < \infty\} = \{0 < \gamma_A < \infty\}.$$

The final result is, for each Borel set $A \subset \overline{E}$ and each $x \in X$:

$$\mathbb{P}_x(x_{\gamma_A^-} \in B) = L^A(x, B) = \int_B u(x, y)\mu_A(dy);$$

in particular,

$$\mathbb{P}_x(T_A < \infty) = L^A(x, \overline{A}) = \int_{\overline{A}} u(x, y)\mu_A(dy). \tag{6.14}$$

Theorem 6.12 *Suppose that* M *satisfies Assumptions* 6.1 *and* 6.2. *Let* $x_0 \in X$ *be the initial state. For any closed set* A *of* X *we have*

$$\mathbb{P}_{x_0}(T_A < \infty) \leq Cap_K(A) \tag{6.15}$$

where Cap_K *is the capacity[2] defined, using* (6.8), *with respect to the Martin kernel* K *defined by* (6.13).

---

[2]This capacity is called Martin capacity.

Proof To bound from above the probability of ever hitting A, consider the hitting time T_A and the last exit time γ_A of A, and the distribution

$$\nu_x(\Lambda) = L^A(x, \Lambda) = \mathbb{P}_x(0 < \gamma_A | x_{\gamma_A -} \in \Lambda); \quad \Lambda \in \mathscr{B}(\mathbf{X}).$$

The Chung's result says that [53]

$$L^A(x, \Lambda) = \int_\Lambda u(x, y) \mu_A(dy); \quad \Lambda \in \mathscr{B}(\mathbf{X}),$$

where μ_A is the equilibrium measure of A, which is given by

$$\mu_A(dy) = L^A(x, dy)u(x, y)^{-1} = \nu_x(dy)u(x, y)^{-1}, \quad \forall x \in \mathbf{X};$$

in particular, for the initial state $x_0 \in \mathbf{X}$,

$$\mu_A(dy) = L^A(x_0, dy)u(x_0, y)^{-1} = \nu_{x_0}(dy)u(x_0, y)^{-1}.$$

It follows that

$$\int_A K(x, y)\nu_{x_0}(dy) = \int_A K(x, y)u(x_0, y)\mu_A(dy)$$

$$= \int_A u(x, y)\mu_A(dy) = \mathbb{P}_x(T_A < \infty) \le 1. \quad (6.16)$$

Therefore, $K(\nu_{x_0}, \nu_{x_0}) \le \nu_{x_0}(A)$ and thus

$$\mathrm{Cap}_K(A) \ge \left[K\left(\frac{\nu_{x_0}}{\nu_{x_0}(A)} \right) \right]^{-1} \ge \nu_{x_0}(A),$$

which by (6.16) yields the upper bound on the probability of hitting A. $\square$

Let A be a closed set of $\mathbf{X}$ and let

$$\mathbb{P}(x, A) = \mathbb{P}_x\big(Reach_\infty(A)\big).$$

For any $x \in \mathbf{X}$ we define the *harmonic measure* of A as follows:

$$H^A(x, B) = P^x(x_{T_A} \in B), \quad B \in \mathscr{B}(\mathbf{X}).$$

Clearly, the support of $H^A(x, \cdot)$ is A and

$$\mathbb{P}(x, A) = H^A(x, A).$$

A weaker result for the hitting probability estimation can be obtained starting with the following proposition:

Proposition 6.13 [153] *If the given process $\{x_t, t \ge 0\}$ has the strong Markov property then, for any transient closed set $A \in \mathscr{B}(\mathbf{X})$ and for $y \in A$ and $x \notin A$, the following equality holds:*

$$u(x, y) = \int_A u(z, y)H^A(x, dz).$$

The hitting probability estimation follows as

Corollary 6.14 *Under the above hypotheses,*

$$\mathbb{P}(x, A) \leq \frac{u(x, y)}{\inf_{z \in A} u(z, y)}, \qquad \forall y \in A.$$

Proof

$$u(x, y) = \int_A u(z, y) H^A(x, dz)$$
$$\geq \inf_{u \in A} u(u, y) H^A(x, A)$$
$$= \mathbb{P}(x, A) \inf_{z \in A} u(z, \beta).$$

Thus,

$$\mathbb{P}(x, A) \leq \frac{u(x, y)}{\inf_{z \in A} u(z, y)}, \qquad \forall y \in A. \qquad \square$$

6.5 Reachability Estimation via Martingale Theory

In this section, reachability questions are characterised as hitting time problems. This section is based on our paper [42]. Let us consider a fixed start point $x_0 \in \mathbf{X}$ and the sequence

$$T_1 < T_2 < T_3 < \cdots$$

of jump times associated to the realisation of a stochastic hybrid system H starting in x_0.

Let A be a Borel subset of $\mathbf{X}$. Let $\mathscr{L}$ be the generator of the Markov process $\mathbf{M}$ corresponding to H.

6.5.1 Method Based on the Martingale Problem

We denote by $D(\mathscr{L})$ the domain of the infinitesimal generator of $\mathbf{M}$. It is known (see Chap. 2) that the process $(C_t^u)_{t \in \mathbb{R}_+}$ defined by

$$C_t^u = u(x_t) - u(x_0) - \int_0^t \mathscr{L}u(x_s) \, ds$$

is a martingale for each u in the domain $D(\mathscr{L})$. This fact implies that for each $t > 0$ we have

$$\mathbb{E}\left[u(x_{t \wedge T_A}) - u(x_0) - \int_0^{t \wedge T_A} \mathscr{L}u(x_s) \, ds \right] = 0.$$

Since we supposed that $T_A < \infty$ a.s., letting $t \to \infty$ gives

$$\mathbb{E}\big[u(x_{T_A})\big] - \mathbb{E}\big[u(x_0)\big] - \mathbb{E}\bigg[\int_0^{T_A} \mathscr{L}\big(u(x_s)\big)\bigg] ds = 0. \tag{6.17}$$

We consider now, as we did in Chap. 5 (see, also, [113]), the *occupation measure* μ_0 and *hitting distribution* μ_1 by

$$\mu_0(B) = \mathbb{E}\bigg[\int_0^{T_A} 1_B(x_s)\,ds\bigg]$$

and

$$\mu_1(B) := H^A(B) = \mathbb{P}(x_{T_A} \in B)$$

for all $B \in \mathscr{B}(\mathbf{X})$.

It is clear that if $B \subset A$ then $\mu_0(B) = 0$ and if $B \subset A^c$ then $\mu_1(B) = 0$. Therefore, μ_0 is concentrated in A^c and μ_1 is concentrated in A. With an integrability argument one can obtain from (6.17) the following so-called *adjoint equation*:

$$\int_{A^c} \mathscr{L}u(x)\mu_0(dx) + u(x_0) - \int_A u(x)\mu_1(dx) = 0. \tag{6.18}$$

Also, it is clear that

$$\mu_0\big(A^c\big) = \mathbb{E}(T_A) \quad \text{and} \quad \mu_1\big(A^c\big) = 0. \tag{6.19}$$

If we take $u = I_{A^c}$ and we suppose that $x_0 \notin A$, then the adjoint equation becomes

$$\int_{A^c} \mathscr{L}I_{A^c}(x)\mu_0(dx) + 1 = 0.$$

It is known that for every $u \in D(\mathscr{L})$, $\mathscr{L}u$ is given by

$$\mathscr{L}u(x) = \mathscr{L}_{\mathrm{cont}}u(x) + \lambda(x)\int_{\mathbf{X}}\big(u(y) - u(x)\big)R(x, dy). \tag{6.20}$$

Thus, we obtain

$$\int_{A^c}\bigg[\lambda(x)\int_{\mathbf{X}}\big(1_{A^c}(y) - 1_{A^c}(x)\big)R(x, dy)\bigg]\mu_0(dx) + 1 = 0,$$

and, if we denote

$$\Lambda_A(x) := \lambda(x)\int_A R(x, dy) = \lambda(x)R(x, A),$$

then

$$\int_{A^c}\Lambda_A(x)\mu_0(dx) = 1.$$

Thus, once the measure μ_0 has been determined (using, e.g., linear programming methods), the hitting time mean can be obtained from (6.19).

6.5.2 Method Based on Martingale Inequalities

The stochastic reachability problem can be situated at the intersection of three disciplines: Markov processes, potential theory and stochastic optimal control. The main engine for moving back and forth between these is provided by (sub, super) martingales. The connections between potential theory and probability theory are now well studied with more than fifty years of history, originating in the work of Doob [73, 74] and Hunt [128–130]. Doob showed how various classical potential theory concepts correspond to properties of superharmonic functions on Brownian motion paths. Hunt extended the ideas to found what is now called *probability potential theory*, in which each of a large class of Markov processes corresponds to a potential theory and conversely.

Concretely, it is known that if $(W_t)_{t \geq 0}$ is the Brownian motion in $\mathbb{R}^d$ and u is a positive superharmonic function in $\mathbb{R}^d$, the $u(W_t)$ is a càdlàg supermartingale in $[0, \infty]$, provided that $\mathbb{E}[u(W_0)] < \infty$. This result was proved in Doob's 1955 paper on the heat equation. It is a major recourse in Hunt's theory, where u became an 'excessive' function and the Brownian motion became a Hunt process. In a general way, Doob's work on (sub, super) harmonic functions with the vital underpinning by Brownian paths, paved the way for Hunt's theory on Markov processes and potentials [128–130]. In the last half of the twentieth century, Hunt's theory was developed further for the standard Markov processes and then for right Markov processes (which might be thought of as the modern generalisations of Hunt processes) by different authors (see, for example, [67]).

On the other hand, martingales represent now a versatile tool for studying different problems related to Markov processes (convergence theorems, invariance principles, etc.). As well, it is now common practice to study stochastic optimal control problems via martingale methods.

For a right Markov process $(x_t)_{t \geq 0}$ (in our case, the realisation of the stochastic hybrid automaton H) and an excessive function u, it is known that the process $(u(x_t))_{t \geq 0}$ is a right continuous $\mathbb{P}_x$-supermartingale for any $x \in \mathbf{X}$ such that $u(x) < \infty$ [67]. Then, for the purpose of this section, we need to review the following supermartingale inequalities (some background on supermartingales is given in the Appendix).

Theorem 6.15 Supermartingale inequalities *Let* $\{Y_t\}_{t \geq 0}$ *be a real-valued supermartingale. Let* $[a, b]$ *be a bounded interval in* $\mathbb{R}_+$. *Then*

$$
\begin{aligned}
c\mathbb{P}\left\{\omega \,\middle|\, \sup_{a \leq t \leq b} Y_t(\omega) \geq c\right\} &\leq \mathbb{E}Y_a + \mathbb{E}Y_b^- \\
c\mathbb{P}\left\{\omega \,\middle|\, \inf_{a \leq t \leq b} Y_t(\omega) \leq -c\right\} &\leq \mathbb{E}Y_b^-
\end{aligned}
\tag{6.21}
$$

hold for all $c > 0$.

Target Sets as Level Sets   Suppose that the target set A in the state space is described as a level set for a given function $F : X \to \mathbb{R}$, i.e.,

$$A = \{x \in \mathbf{X} \,|\, F(x) \geq l\}.$$

F can be chosen, for example, to be the Euclidean norm or the distance to the boundary of E. The probability of the set of trajectories that hit A until time horizon $T > 0$ can be expressed as

$$\mathbb{P}\left\{ \sup_{t \in [0,T]} F(x_t) \geq l \right\}. \tag{6.22}$$

Our main goal is to study the stochastic process $(F(x_t))_{t \geq 0}$ and suitable hypotheses for F such that upper bounds for the probabilities (6.22) can be easily derived.

$F(x_t)$ represents the best candidate for defining a possible abstraction for $\mathbf{M}$, which preserves the reach set probabilities. The main difficulty is that $F(x_t)$ is a Markov process only for special choices of F.

In this section, we propose further to 'approximate' the function F such that the stochastic process $(F(x_t))_{t \geq 0}$ becomes a supermartingale. Then some upper bounds for (6.22) can be easily derived from the martingale's inequalities that are well studied in the literature. In fact, our aim is to find an excessive function (or stochastic Lyapunov function) $\widetilde{F}$ close enough to F so that different properties of the supermartingale $(\widetilde{F}(x_t))_{t \geq 0}$ can be exploited.

The main goal of this section is to find upper bounds for probabilities that appear in the stochastic reachability problem, using the theory of (sub/super) martingales associated to the realisation of a stochastic hybrid system.

Methodology Description   Suppose now that $A \in \mathscr{B}(\mathbf{X})$ is a target set described as a level set associated to a *positive measurable function* $F : \mathbf{X} \to \mathbb{R}_+$. If F is an excessive function then the problem of finding an upper bound for reach set probability (6.22) is easily solved by (6.21) since $(F(x_t))$ is a $\mathbb{P}_x$-supermartingale for any $x \in \mathbf{X}$ such that $F(x) < \infty$.

The function F might have different shapes and in order to allow more freedom in choosing it, we *cannot* suppose a priori that F is an excessive function. Also, it might be difficult to check its excessiveness using the infinitesimal generator $\mathscr{L}$, i.e., F satisfies the condition

$$\mathscr{L}F \leq 0.$$

Even worse, it could happen that $F \notin D(\mathscr{L})$.

In the following, we use an 'approximation' of function F by an excessive function. The best candidate for such an approximation is the réduite $\mathbf{R}F$. Let us denote $\mathbf{R}F$ by $\widetilde{F}$. Since $\widetilde{F}$ is an excessive function, the following result, which gives an upper bound of the reach set probabilities, is just a consequence of Theorem 6.15.

Theorem 6.16 *For any $x \in \mathbf{X}$ such that $\widetilde{F}(x) < \infty$, we have*

$$\mathbb{P}_x\left\{ \sup_{t \in [0,T]} \widetilde{F}(x_t) \geq l \right\} \leq \frac{\widetilde{F}(x) + \lim_{t \nearrow T} \mathbf{P}_t \widetilde{F}(x)^-}{l}, \qquad l > 0 \tag{6.23}$$

where $(\mathbf{P}_t)$ is the operator semigroup.

In Theorem 6.16, we have used the fact that $\mathbf{P}_0$ is the identity operator.

Moreover, the inequality (6.23) can be written in terms of the infinitesimal generator $\mathscr{L}$ of $\mathbf{M}$. This fact is possible because there exists a classical result [83], which says that the semigroup $(\mathbf{P}_t)_{t>0}$ formula can be computed using the infinitesimal generator $\mathscr{L}$ expression via the following connection:

$$\mathbf{P}_t = \exp(t\mathscr{L}), \quad t > 0.$$

6.5.3 Method Based on the Barrier Certificates

In the previous subsection, we have obtained some upper bounds for the reach probabilities, as a consequence of the Doob Inequality. This inequality has inspired the concept of *barrier certificate* for stochastic safety verification [194]. This stochastic barrier certificate is a supermartingale associated to the dynamics of a given system. In addition, this barrier certificate is required to have a value for the initial state, which is lower than its values on the unsafe region. Then again the Doob Inequality is employed to obtain upper bounds for the reach probability of the unsafe region.

First, we will explain this method for diffusion processes, then for switching diffusions and at the end for other classes of stochastic hybrid processes.

Suppose that (x_t) is a diffusion process that is defined as the strong solution for the SDE

$$dx_t = b(x_t) + \sigma(x_t)\,dW_t, \quad x_0 = \mathbf{x}_0 \tag{6.24}$$

where $b : \mathbb{R}^+ \times \mathbb{R}^d \to \mathbb{R}^d$ and $\sigma : \mathbb{R}^+ \times \mathbb{R}^d \to \mathbb{R}^{d \times m}$ are measurable functions (vector- and matrix-valued, respectively), (W_t) an m-dimensional Wiener process and $\mathbf{x}_0 \in \mathrm{Init} \subseteq \mathbb{R}^d$. Assume that the conditions of Theorem 2.2 for the existence and uniqueness of the solution of this SDE are fulfilled. Denote by L the infinitesimal generator of (x_t) given by the formula (2.29). We can restrict this diffusion to a smaller domain $\mathbf{X} \subseteq \mathbb{R}^d$ considering that

$$x_t = \Delta, \quad \forall t > \tau_{\mathbf{X}},$$

where Δ is a cemetery point and $\tau_{\mathbf{X}}$ is the first exit time from $\mathbf{X}$ (or the first hitting time of $\mathbb{R}^d \backslash \mathbf{X}$).

Let us consider that F is a nonnegative excessive function for the diffusion process on $\mathbf{X}$. Using the background provided in the previous subsection, we find that $F(x_t)$ is a supermartingale, i.e., its expected value is nonincreasing with time. Then, using Doob's inequality, we have

$$\mathbb{P}_x \left\{ \sup_{0 \leq t < \infty} F(x_t) \geq \lambda \right\} \leq \frac{F(x)}{\lambda}$$

for any initial condition $x \in \mathbf{X}$ for (x_t).

As usual, let $A \in \mathscr{B}(\mathbf{X})$ be the target set for the stochastic reachability analysis, which is considered an unsafe set in this context.

Theorem 6.17 [194] *Suppose that there exists a twice continuously differentiable $F : \mathbf{X} \to \mathbb{R}_+$ such that*

$$F(x) \geq 1, \quad \forall x \in A, \tag{6.25}$$

$$F(x) \leq \gamma, \quad \forall x \in \text{Init}, \; \gamma \leq 1, \tag{6.26}$$

$$LF \leq 0, \quad \forall x \in \mathbf{X}. \tag{6.27}$$

Then

$$\mathbb{P}_x\{\omega | \exists t \geq 0 \text{ such that } x_t \in A\} \leq \gamma$$

for any initial condition $x \in \text{Init}$.

Proof The proof is rather simple and is based on the properties of the excessive functions. The inequality (6.27), which is the characterisation of the excessive functions for a stochastic process, implies that $F(x_t)$ is a supermartingale (by using the Dynkin formula (2.35)). Then, for any initial condition $x \in \text{Init}$, using the Doob Inequality, we have

$$\mathbb{P}_x\{\omega | \exists t \geq 0 \text{ such that } x_t \in A\} \leq \mathbb{P}_x\left\{ \sup_{0 \leq t < \infty} F(x_t) \geq 1 \right\}$$

$$\leq F(x) \leq \gamma. \qquad \square$$

Reference [194] provides methods for the computation of an upper bound γ and the barrier certificate F for stochastic polynomial systems via a sum of squares decomposition and semidefinite programming, using the software SOSTOOLS [195].

The straightforward generalisation of this method can be performed in the first step for switching diffusion processes. Suppose that

$$H = (Q, X, u, \mu_0, \sigma, \lambda_{ij})$$

is a SDP with the state space $\mathbf{X}$ (see the definition in Sect. 3.3.4). Here, $\mu_0 = (\mu_{0,q})_{q \in Q}$ is the initial probability measure. It is known that the realisation of an SDP is a strong Markov process $\mathbf{M} = (x_t)$, whose trajectories are continuous on $[0, \infty)$. The hybrid state space is

$$\mathbf{X} := Q \times X.$$

For an SDP, the infinitesimal generator is a generalisation of a diffusion generator and can be expressed as follows (see, for example, [105]):

$$\mathscr{L}f(q, z) = \sum_{i=1}^{n} u_i(q, z) \frac{\partial f(q, z)}{\partial z_i}$$

$$+ \frac{1}{2} \sum_{i,j=1}^{n} [\sigma(q, z)\sigma(q, z)^{\mathsf{T}}]_{ij} \frac{\partial^2 f(q, z)}{\partial z_i \partial z_j}$$

$$+ \sum_{k=1}^{N} \lambda_{qk}(z) f(k, z). \tag{6.28}$$

This generator maps functions twice differentiable on **X** with compact support to continuous functions with compact support on the same state space.

For this class of stochastic hybrid systems, a barrier certificate can be constructed by 'piecing out' some barrier certificates for the component diffusions. This means that

$$F(q, z) := F_q(z), \quad q \in Q$$

where each F_q is a barrier certificate for the diffusion evolving in the location q. Of course, F should satisfy a global condition with respect to the entire SDP.

Let $A \in \mathscr{B}(\mathbf{X})$ be a target set. Additionally, suppose that the initial state of the SDP is chosen from a set $X_0 \subset \mathbf{X}$. Let us denote

$$X_{0,q} := \{z \in X | (q, z) \in X_0\},$$
$$A_q := \{z \in X | (q, z) \in A\}.$$

The main result can be formulated as follows.

Theorem 6.18 [194] *Let* **M** *be the realisation of a SDP H. Suppose that there exists a collection of twice differentiable nonnegative functions* $(F_q)_{q \in Q}$ *such that*

$$F_q(z) \geq 1, \quad \forall z \in A_q;$$
$$\mathscr{L}F(q, z) \leq 0, \quad \forall z \in X;$$

and in addition

$$\sum_{q \in Q} \int_{X_{0,q}} F_q(z) \, d\mu_{0,q}(z) \leq \gamma.$$

Then

$$\mathbb{P}_{\mu_0}\{\omega | \exists t \geq 0 \text{ such that } x_t \in A |\} \leq \gamma.$$

Proof The proof is similar to the proof of Theorem 6.17. The only difference is that we have to use the expression of

$$\mathbb{P}_{\mu_0}(E) = \int \mathbb{P}_x(E) \, d\mu_0(x). \qquad \square$$

It is clear that Theorem 6.18 remains true for more general classes of stochastic hybrid systems like GSHSs or PDMPs. The proof does not require continuity of the paths of the underlying stochastic hybrid process. The main assumption is that the barrier certificate F should be found in the domain of the generator $\mathscr{L}$. Moreover, it is clear that the exact expression of this generator does not play any role in the proof. The main difficulty one encounters when dealing with GSHSs, which allow forced discrete transitions, is that the computation of barrier certificates can no longer be based on SOSTOOLS.

6.6 Reachability Estimation via Quadratic Forms

This section is developed using our paper [43]. The basic idea of the reachability method proposed here is to employ the characterisation of the strong Markov processes based on the associated quadratic forms, called *Dirichlet forms*, defined using the process generator (see Chap. 2). A Dirichlet form makes possible the use of the operator theory for a given Markov process. Also, it comes with a specific imprecise probability, called *capacity associated to the Dirichlet form*. This capacity can be expressed in terms of the hitting times of the corresponding Markov process [176]. We investigate the possible benefits of applying a Dirichlet form-based method to the study of the reachability problem of stochastic hybrid systems.

We briefly explain the methodology to obtain upper bounds for the reach set probabilities. First, we suppose that the target sets in the state space of a GSHS H are given as level sets. Then we start with a Borel set A specified as a level set for a smooth function F. The Markovian properties of the GSHS realisation **M** allow us to define the corresponding quadratic form $\mathscr{E}$. This form is a special kind of Dirichlet form (i.e., it satisfies the special axioms that characterise the Dirichlet forms corresponding to Borel right processes).

For a detailed treatment of general Dirichlet forms, the reader is referred to [176]. We consider that for the reachability problem studied in this section, it is not relevant to present the whole mathematical apparatus which characterises Dirichlet forms. Intuitively, Dirichlet forms are quadratic forms that encode the Markovian properties of the underlying process. The relevance of this theory in our study is that we can obtain results on the estimation of reachability probabilities using properties of the capacity associated to a Dirichlet form.

Secondly, using the function F we define the *induced process* $F(x_t)$, which is a one-dimensional stochastic process (with values in $\mathbb{R}$) and the induced Dirichlet form $\mathscr{E}^*$ (see formula (6.30) below). The form $\mathscr{E}^*$ is associated with $F(x_t)$ if the induced process is Markov.

The mechanism used to obtain the upper bounds for the reach set probabilities is based on the inequality that can be derived between the capacity of the initial Dirichlet form $\mathscr{E}$ and the capacity corresponding to the induced Dirichlet form $\mathscr{E}^*$ (the inequality (6.34) below). The relation existing between hitting times and capacities allows us to obtain appropriate upper bounds for the reach probabilities.

6.6.1 Target Sets

Usually, a target set A in the state space would be represented as a *level set* for a given function $F : \mathbf{X} \to \mathbb{R}$, i.e.,

$$A = \left\{ x \in \mathbf{X} \,\middle|\, F(x) > l \right\}.$$

The function F can be chosen suitably with respect to the practical needs. For example, F could be the Euclidean norm or the distance to the boundary of A. The probability of the set of trajectories, which hit A over time horizon $T > 0$, can be expressed as (6.22).

6.6.2 Dirichlet Forms

The strong Markov property of GSHS allows us to define an associated quadratic form, as follows.

Let H be a GSHS whose realisation is described by a Markov process $\mathbf{M}$. Let $D(\mathcal{L})$ be the domain (not the extended domain) of its infinitesimal generator $\mathcal{L}$. Using the Lebesgue measure λ^i on $\mathbb{R}^{d(i)}$, we define a new measure μ on the state such that for each $i \in Q$ the projection of μ to each mode X^i is exactly the restriction of the Lebesgue measure λ^i to that mode, $\lambda^i|_{X^i}$, i.e.,

$$\mu(\{i\} \times A) := \lambda^i(A), \quad A \in \mathcal{B}(\mathbb{R}^{d(i)}).$$

Let μ^* be the image of μ through the map F.

Remark 6.19 Since any GSHS provides a Borel right process, the process semigroup $(\mathbf{P}_t)$ may be viewed as a strongly continuous semigroup of operators on $L^2(\mathbf{X}, \mu)$ [176]. Its generator is defined by the same limit

$$\lim_{t \searrow 0} \frac{1}{t}(\mathbf{P}_t f - f) \qquad (6.29)$$

with respect to the norm of $L^2(\mathbf{X}, \mu)$. The domain of the generator consists of those $f \in L^2(\mathbf{X}, \mu)$ for which the limit (6.29) exists in the norm of $L^2(\mathbf{X}, \mu)$.

Remark 6.20 [176] Under the standard assumptions regarding H, there exists a *quasi-regular Dirichlet form*[3] $(\mathcal{E}, D[\mathcal{E}])$ on $L^2(\mathbf{X}, \mu)$ associated with the process (x_t), given by

$$\begin{cases} D(L) \subset D[\mathcal{E}] \\ \mathcal{E}(u, v) = \langle -Lu, v \rangle, \quad u \in D(L), \quad v \in D[\mathcal{E}]. \end{cases}$$

We can think of a Dirichlet form $\mathcal{E}$ as a recipe for a Markov process $(x_t)_{t \geq 0}$, in the sense that $\mathcal{E}$ describes the behaviour of the composed process $u(x_t)$ for every u in the domain of $\mathcal{E}$. There is no guarantee that the 'coordinates' $(u(x_t))_u$ can be put together in a consistent way to form a process with reasonable sample paths.

6.6.3 Induced Dirichlet Forms

Let us denote the sub-σ-algebra of $\mathcal{B}$, generated by $\sigma(F)$, and the projection operator from $L^2(\mathbf{X}, \mathcal{B}, \mu)$ to $L^2(\mathbf{X}, \sigma(F), \mu)$ by $\mathcal{F}$. In the case when μ is a probability measure, we have

$$\mathcal{F} \equiv \mathbb{E}_\mu[\cdot|F].$$

---

[3] See Definition 3.1 from [176].

More precisely, the function F induces a form $\mathscr{E}^*$ on $L^2(\mathbb{R}, \mu^*)$ by

$$\mathscr{E}^*\big(u^*, v^*\big) = \mathscr{E}\big(u^* \circ F, v^* \circ F\big); \quad u^*, v^* \in D\big[\mathscr{E}^*\big] \tag{6.30}$$

where

$$D\big[\mathscr{E}^*\big] = \big\{u^* \in L^2(\mathbb{R}, \mu^*) \big| u^* \circ F \in D[\mathscr{E}]\big\}.$$

Proposition 6.21 [135] *If $\mathscr{F}(\mathscr{D}) \subset D[\mathscr{E}]$, where $\mathscr{D}$ is some L^2-dense subset of $D[\mathscr{E}]$, then $\mathscr{E}^*$ is a Dirichlet form on $L^2(\mathbb{R}, \mu^*)$.*

Assumption 6.3 Suppose that the Dirichlet form $\mathscr{E}^*$ is quasi-regular.[4]

In [135], it is shown that, under a mild condition on the function F, Assumption 6.3 can be accomplished. This assumption ensures that there exists (x_t^*), which is a right Markov process with the state space $\mathbb{R}$, associated with the Dirichlet form $\mathscr{E}^*$ [5]. If $F(x_t)$ happens to be Markovian then $\mathscr{E}^*$ is its associated Dirichlet form (see [200] for conditions on F that imply the Markov property of $F(x_t)$).

Assumption 6.4 Suppose that the Dirichlet forms $\mathscr{E}$, $\mathscr{E}^*$ are symmetric[5]

$$\mathscr{E}(u, v) = \mathscr{E}(v, u), \quad u, v \in D[\mathscr{E}],$$
$$\mathscr{E}^*\big(u^*, v^*\big) = \mathscr{E}^*\big(v^*, u^*\big), \quad u^*, v^* \in D\big[\mathscr{E}^*\big].$$

Assumption 6.4 is not restrictive (any result valid for regular Dirichlet forms and invariant under quasi-homeomorphisms is applicable to quasi-regular Dirichlet forms [48]).

Each (quasi-regular) symmetric Dirichlet form can be expressed as the sum of its parts—continuous, jumping and killing—corresponding to the same parts of the Markov process considered. Precisely, a regular Dirichlet form $\mathscr{E}$ can be decomposed using *the Beurling–Deny representation* [93]:

$$\mathscr{E}(u, v) = \mathscr{E}_c(u, v) + \int_{\mathbf{X} \times \mathbf{X} \backslash d} \big[u(x) - u(y)\big]\big[v(x) - v(y)\big] J(dx, dy)$$
$$+ \int_{\mathbf{X}} u(x)v(x)k(dx), \tag{6.31}$$

$u, v \in D[\mathscr{E}] \cap C_0(\mathbf{X})$. Here, $\mathscr{E}_c$ is a symmetric form with domain $D[\mathscr{E}_c] = D[\mathscr{E}]$, which satisfies the property

* $\mathscr{E}_c(u, v) = 0$ if $u, v \in D[\mathscr{E}]$ have support compact and v is constant on a neighbourhood of $\text{supp}[u]$.

---

[4]See Definition 3.1 from [176].

[5]See [93, 176] for the theory of symmetric and nonsymmetric Dirichlet forms.

J is a symmetric positive measure on $\mathbf{X} \times \mathbf{X}\backslash d$, d being the diagonal; and k is a positive measure on $\mathbf{X}$.

The form $\mathscr{E}_c$ and measures J and k are uniquely determined by $\mathscr{E}$. $\mathscr{E}_c$ is called *the diffusion part* of $\mathscr{E}$ and J and k are called the *jump measure* and the *killing measure*, respectively, associated with $\mathscr{E}$.

Let us make the following notations:

$$\mathscr{E}(u) := \mathscr{E}(u, u); \quad u \in D[\mathscr{E}],$$
$$\mathscr{E}^*(u) := \mathscr{E}^*(u^*, u^*); \quad u^* \in D[\mathscr{E}^*],$$
$$\|u\|_\mu^2 := \langle u, u \rangle_\mu,$$
$$\|u^*\|_{\mu^*}^2 := \langle u^*, u^* \rangle_{\mu^*},$$

where $\langle \cdot, \cdot \rangle_\mu$ is the inner product of $L^2(\mathbf{X}, \mu)$ and $\langle \cdot, \cdot \rangle_{\mu^*}$ is the inner product of $L^2(\mathbb{R}, \mu^*)$.

Note if A^* is an open set in $\mathbb{R}$ and

$$A := F^{-1}(A^*)$$

then we can define for $p > 0$, the *p-capacity* of A

$$\mathrm{Cap}_p(A) = \inf\{\mathscr{E}(u) + p\|u\|_\mu^2 \,|\, u \in D[\mathscr{E}], u \geq 1 \ \mu\text{-a.e. on } A\} \qquad (6.32)$$

and the *p-capacity* of A^*

$$\mathrm{Cap}_p^*(A^*) = \inf\{\mathscr{E}^*(u) + p\|u^*\|_{\mu^*}^2 \,|\, u^* \in D[\mathscr{E}^*], u^* \geq 1 \ \mu^*\text{-a.e. on } A^*\}, \qquad (6.33)$$

where μ-a.e. (μ almost everywhere, i.e., with exception of a μ-negligible set) means outside of a set with μ-measure zero.

Proposition 6.22 [135] *Under Assumptions* 6.3 *and* 6.4, *if* A^* *is open and* $A = F^{-1}(A^*)$ *then*

$$\mathrm{Cap}_p(A) \leq \mathrm{Cap}_p^*(A^*). \qquad (6.34)$$

We can consider the two first hitting times T_A (with respect to (x_t)) and T_{A^*} (with respect to (x_t^*)). Intuitively, the capacity (6.32) (respectively (6.33)) is the Laplace transform of the expectations of hitting time T_A (respectively T_{A^*}) of the target set (respectively of the 'induced' target set).

6.6.4 Upper Bounds for Reach Set Probabilities

An upper estimation for the reach probabilities will be given in terms of the Dirichlet form induced by F on $\mathbb{R}$. This form corresponds to the process $F(x_t)$.

Assumption 6.5 Assume that $\mu(\mathbf{X}) < \infty$, $1 \in D[\mathscr{E}]$ and $k(\mathbf{X}) < \infty$, where the killing measure k is described in the Beurling–Deny representation (6.31).

The translation of the capacitary inequality (6.34) into probabilistic terms for the right Markov processes (x_t) and (x_t^*) associated with $\mathscr{E}$ and $\mathscr{E}^*$ gives rise to the following result:

Proposition 6.23 *Suppose that Assumption 6.5 is fulfilled. If $A^* \subset \mathbb{R}$ is an open set of finite Cap^*-capacity and $A = F^{-1}(A)$ then for all $p > 0$, $T > 0$, by rescaling the horizon time $T := T_p \cdot p$, we have*

$$\mathbb{P}_\mu(T_A \leq T) \leq e^p \{ \mathbb{E}_{\mu^*} \exp(-T_p^{-1} T_{A^*}) + T_p \mathbb{E}_{k^*} \exp(-T_p^{-1} T_{A^*}) \},$$

where k^* is the killing measure associated with the killing part of $\mathscr{E}^*$. Also

$$\mathbb{P}_\mu(T_A \leq T) \leq T_p e^p \mathrm{Cap}_{T^{-1}}(A^*) \tag{6.35}$$

where

$$\mathrm{Cap}_{T^{-1}}(A^*) = \inf \{ \mathscr{E}^*(u) + T^{-1} \|u^*\|_{\mu^*}^2 \, | \, u^* \in D[\mathscr{E}^*], u^* \geq 1, \mu^*\text{-a.e. on } A^* \}.$$

One might, for instance, use the small induced processes rather than the huge original process to deal with the reachability problem. The induced Dirichlet form capacity (of $A^* = (l, \infty)$) plays an essential role in obtaining the reach event probability estimation. If the model H is discretised then the induced process is a one-dimensional jump process and, therefore, the computation of Laplace transform and the mean level-crossing time is feasible. It is interesting to note that the capacity of the target set is subadditive. So even if the target set were very complex, then the capacity of the target set is at most the sum of capacities of its parts.

6.7 Some Remarks

This chapter is the result of different probabilistic approaches on stochastic reachability. Starting with the characterisations given in the previous chapter, we attacked the stochastic reachability problem only from the perspective of applied probabilities. Undoubtedly, we need to make an important point: The measures involved in stochastic reachability analysis are indeed closely related to some important concepts and facts that characterise Markov processes (transition semigroup, resolvent, generator, forward/backward Kolmogorov equations, capacities), but these connections only draw possible research directions. The links between stochastic reachability and specific concepts associated to a Markov process do not solve our problem. First, because sometimes, even for the simpler cases (as for Markov chains or diffusion processes), the problem does not admit straightforward solutions with computational methods associated. Second, even if we have nice methods and solutions for the simpler cases, these cannot be easily extended to the framework of hybrid systems. Ideally, suitable methods for stochastic reachability of hybrid systems should combine, in a cross-augmenting manner, reasonably good existing methods for discrete processes with methods that admit computational tools for continuous processes.

We can remark that for discrete space processes, we have multiple choices for solving stochastic reachability in an exact way. But, for continuous (diffusion) processes, the only method that provides an exact solution is based on PDEs. Solving these PDEs is a challenge even for the case when the dimension of the state space is greater than one. If we try to combine these types of method (for discrete and continuous cases), we might get in trouble. Definitely, we will have to solve integro-differential equations! In order to solve such equations, we have to leave the strictly probabilistic framework.

Therefore, the reasonable conclusion is that if we want to employ purely probabilistic methods for stochastic reachability, the best results of such an approach would consist of upper/lower bounds for the reach probabilities. The probabilistic mechanism works very well for discrete space processes, but it should be enhanced with PDEs or functional analysis techniques when the continuous space features are added.

Chapter 7
Analytic Methods for Stochastic Reachability

7.1 Overview

This chapter is dedicated to the investigations of the analytic solutions of stochastic reachability for hybrid systems. Based on the connections between the stochastic reachability problem and other well-studied stochastic control problems (like the optimal stopping, obstacle problem, exit problem) we derive appropriate analytic methods for the computation of the reach probability characterised as a value function of such control problems. We do not get these methods for free. Since the underlying stochastic process is hybrid, it has complex dynamics and guards that govern the discrete transitions. Therefore, these methods have to be carefully reconsidered and adapted and sometimes new innovative solutions have to be discovered. The interaction between the continuous evolutions and the jumping mechanism of a stochastic hybrid process has a big impact in the study of reachability issues. This interaction should be investigated in a critical way so consequences regarding undesirable properties of the different control value functions can be overcome.

7.2 Réduite and Optimal Stopping

In this section, we characterise the reach set probabilities using the concepts of *balayage* and *réduite* from potential theory and the concept of *hitting operator* from Markov process theory. Let us review the following result.

Proposition 7.1 *For any $x \in \mathbf{X}$ and Borel set $A \in \mathscr{B}(\mathbf{X})$, we have*

$$\mathbb{P}_x\big[Reach_\infty(A)\big] = B_A 1(x) = P_A 1(x) = \mathbb{P}_x[T_A < \zeta].$$

From the fact that on $\mathbf{X} \backslash A$ the equality between balayage and réduite holds,

$$\mathbf{R}_A v = B_A v,$$

L.M. Bujorianu, *Stochastic Reachability Analysis of Hybrid Systems*,
Communications and Control Engineering,
DOI 10.1007/978-1-4471-2795-6_7, © Springer-Verlag London Limited 2012

when the process starts outside of the target set A, finding the reach set probability $\mathbb{P}_x[Reach_\infty(A)]$ is equivalent to finding the réduite $\mathbf{R}(1_A)$. In the next subsection, we make the connection between the concept of réduite and an appropriate *optimal stopping problem* associated to the process.

The existence of the réduite of a bounded measurable function $g : \mathbf{X} \to \mathbb{R}$ (with respect to the resolvent $\mathcal{U}$ associated to a Markov process) can be proved using different arguments. Some probabilistic tools have been used in [79]. In fact, this existence is based on the following equality:

$$\mathbf{R}g(x) = \sup\{\mathbb{E}_x[g(x_S)1_{\{S<\zeta\}}]; S \text{ stopping time}\}. \tag{7.1}$$

The right hand side of equality (7.1) is related with the optimal stopping problem (OSP) associated with a Markov process. For different classes of stochastic processes (diffusions, Feller/Hunt/standard processes), the fact that the optimal value function coincides with the smaller excessive majorant of the exercise payoff is a well-known result. This result has its origin in the famous paper of Dynkin [76], concerning the OSP of a Markov process.

In the following, the optimal stopping problem for a (strong right) Markov process

$$\mathbf{M} = (\Omega, \mathcal{F}, \mathcal{F}_t, x_t, \mathbb{P}_x)$$

taking values in a Lusin space is briefly reviewed. Let Σ denote the set of stopping times (finite or not) with respect to the filtration $\{\mathcal{F}_t\}$, i.e.,

$$\tau \in \Sigma \Leftrightarrow \forall t, \quad \{\tau \le t\} \in \mathcal{F}_t.$$

Consider $g : \mathbf{X} \to \mathbb{R}$ a bounded measurable function called the *reward function* (the interpretation being that if we stop the process at a point $x \in \mathbf{X}$ we obtain a reward $g(x)$). Obviously, the definition of OSP requires some integrability conditions over the paths of $\mathbf{M}$ (see, for example, [79] for more details). Let $(y_t)_{t\ge0}$ be the *reward process* defined by

$$y_t = g(x_t), \quad t \ge 0.$$

The *maximal payoff function* (or the *value function*, in the terminology of [64]) is

$$v(x) := \sup\{\mathbb{E}_x y_\tau | \tau \in \Sigma\}. \tag{7.2}$$

The value function has been characterised in terms of the minimal excessive function lying above the reward function for standard Markov processes [212], or more general for right Markov processes [79]. In the light of the definitions presented here, this means that *the réduite of g coincides with the value function* (7.2).

7.3 Stochastic Reachability as an Optimal Stopping Problem

Let us introduce the *reachability function* $w_A : \mathbf{X} \to [0, 1]$ associated to a measurable set A, defined as

$$w_A(x) := \mathbb{P}_x\big[Reach_\infty(A)\big]. \tag{7.3}$$

Taking the reward function g to be equal with the indicator function of A, i.e.,

$$g := 1_A, \tag{7.4}$$

according to the characterisation of the reach set probability derived in the previous section we obtain the following result:

Proposition 7.2 *If $A \in \mathcal{B}(\mathbf{X})$ then the reachability function w_A coincides with the value function of the reward process $y_t = 1_A(x_t)$, i.e.,*

$$w_A(x) = \sup\big\{\mathbb{P}_x(x_\tau \in A)\big|\tau \in \Sigma\big\}, \quad \forall x \in \mathbf{X}.$$

This is a relevant result vis-a-vis reach set probabilities. Their computation is equivalent to the computation of an OSP value function. The main difficulty is that the reward function g given by (7.4) is not a smooth function; it might be lower/upper semicontinuous according to the topological properties of the target set A. Then standard computational methods for the OSP value functions should be reconsidered when we are dealing with the computation of w_A so that the nonsmoothness of g will not be a big obstacle.

7.4 Optimal Stopping Problem for Borel Right Processes

The realisations of stochastic hybrid systems are (Borel) right processes, and, therefore, the general theory of optimal stopping developed for right processes [25, 180] can be applied. This theory is foundational since it provides mathematical characterisations of the value function using different tools available for right processes:

- The approach presented in [25] relies on a well-known connection between excessivity and a special type of functional concavity originally due to Dynkin and Yushkevich [78].
- The main result of [180] shows that the value function of an optimal stopping problem coincides with *Snell's envelope* of the reward process. Snell's envelope is the smallest supermartingale that dominates the reward process.

Markov processes that appear in the semantics of stochastic hybrid systems are (Borel) right processes, but they may or may not be

- standard Markov processes (whose theory is well developed in [33]), because the quasi-left continuity might fail, due to the existence of the active boundaries when the process jumps to a new mode;
- Feller processes (processes with continuous transition probabilities), since they have predictable jumps (i.e., forced transitions). See [64] for discussion of the Feller property for piecewise deterministic Markov processes.

Therefore, the optimal stopping times do not need to exist, and the treatment of the OSP requires some additional hypotheses. We recall the following inclusions (which are classical in the literature of Markov processes [96]) among the various classes of processes:

$$\{\text{Feller}\} \subset \{\text{Hunt}\} \subset \{\text{special standard}\} \subset \{\text{right}\}.$$

These different processes were introduced at various stages of the recent theory of Markov processes. This leads us to conclude that the well-developed OSP methods available for standard Markov processes [212], or those available for Feller–Markov processes (and their elliptic integro-differential operators) [95] are certainly not directly applicable to the Borel right processes that arise in the GSHS context.

For right Markov processes, the value function has been characterised as the minimal excessive function lying above the reward function [79]. Based on this characterisation, for Borel right processes, computation of the value function corresponding to the OSP must be based on specific features of these processes. Towards this objective, we look at two distinct types of methods:

1. *Analytical methods.* These methods are used when the OSP value function is characterised by:

   - different representations of excessive functions (such as integrals of the Green kernel of the process or the Riesz decomposition [98]);
   - variational inequalities associated to some energy functional constructed using the hitting/balayage operator [99];
   - proving that the réduite (the value function) is the solution of some variational inequality [94, 164], then solving numerically this inequality;

2. *Probabilistic methods.* These include:

   - approximations of the underlying Markov process by a Markov chain and compute the value function corresponding to the chain by some specific algorithms [158, 196];
   - martingale methods based on Snell's envelope;
   - Monte Carlo methods [148].

In the remainder of this chapter, we sketch the first two analytic methods.

7.5 Methods Based on Excessive Function Representations

Methods based on excessive function representations have been exploited for diffusion processes, Hunt processes and Lévy processes [185, 207]. These methods consist of establishing first an integral representation for superharmonic functions and then deriving information about the final behaviour of paths.

In this section, we try to understand the structure of the value function of the OSP in a regular enough framework of Markov processes—precisely the class of Borel right processes. We reach the conclusion that finding the solution of such a problem is equivalent to finding the representation of the value function in terms of the Green kernel. The *support of the measure* that appears in this representation is the *stopping region* for the problem.

We recall that for a Markov process **M** and the optimal stopping problem, one can introduce:

- *continuation set*

$$C = \{x \in \mathbf{X} | v(x) > g(x)\}, \tag{7.5}$$

- and *stopping set*

$$D = \{x \in \mathbf{X} | v(x) = g(x)\}. \tag{7.6}$$

The Riesz decomposition of excessive functions is of key importance in our approach to optimal stopping. Especially in the infinite horizon case, for optimal stopping of diffusion processes, it is seen [207] that this method gives the solution more directly than the much used techniques based on the principle of smooth pasting (the approach of variational inequalities). The Riesz representation is one of the important results in classical theory of partial differential equations. It states that a non-negative superharmonic function in an open set of a Euclidean space can be decomposed uniquely as a sum of a harmonic function and a potential of a measure supported on the given open set (i.e., the integral of the Green kernel with respect to this measure). Under certain duality conditions, Hunt studied the Riesz representation of excessive functions in the setting of Hunt processes. Hunt's result was extended to Borel right processes under weak duality in [98]. A Riesz decomposition of an excessive function was proved in [102] for a Borel right process without duality assumptions.

Using the operator semigroup $\mathscr{P}$ defined by (2.23), one can define the *kernel operator U* by

$$U f(x) = \int_0^\infty \mathbf{P}_t f(x) \, dt, \quad f \in \mathscr{B}^b(\mathbf{X}). \tag{7.7}$$

Uf is called the *potential* of f because, with respect to the generator $\mathscr{L}$ of **M**, the function Uf satisfies the same relation as the classical Newtonian potential with respect to the Laplace operator Δ, i.e., Uf is the solution of the following Poisson equation:

$$-\mathscr{L}\phi = f. \tag{7.8}$$

Of course, if in (7.7), f ranges over the indicator functions of measurable sets, we can write U as a stochastic kernel

$$U(x, A) = \int_0^\infty \mathbb{P}_x(x_t \in A) \, dt.$$

For a random walk, a possible interpretation of $Uf(x)$ according to [78] is as follows: 'Let every hit at the point y bring a payoff $f(y)$. Then $Uf(x)$ is the mean value of the payoff obtained during a random walk of a particle with initial point x.'

We suppose that our process is *transient*, i.e., there is a strictly positive measurable function q such that $Uq \leq 1$. The transience hypothesis guarantees that the cone $\mathcal{E}_{\mathbf{M}}$ is rich enough to be used. For the scope of this section, we can also impose the hypotheses of *weak duality* from [98], or some *analytic conditions,* as in [102].

Suppose that the assumptions from [102] are in force. The main assumption is related to the absolute continuity of the kernel operator U with respect to a σ-finite excessive measure m on $(\mathbf{X}, \mathcal{B})$, called a *reference measure.*

Assumption 7.1 There exists a non-negative $\mathcal{B} \times \mathcal{B}$ measurable function $u(\cdot, \cdot)$ such that

(i) $U(x, dy) = u(x, y)m(dy)$, $x \in \mathbf{X}$;
(ii) $x \mapsto u(x, y)$ is excessive, $y \in \mathbf{X}$.

One needs the *potential density* $u(x, y)$ to define the potential of a measure μ by setting

$$U\mu(x) := \int_{\mathbf{X}} u(x, y)\mu(dy).$$

The potential density plays the role of the Green kernel defined for Brownian motion. For a Markov chain, the Green kernel $u(x, y)$ gives the mean of the sojourn time of chain in state y if it started in x.

Let us introduce briefly a very useful type of function, namely *harmonic functions* for Markov processes. The precise definition of a harmonic function for a right process can be found in [98]. In the theory of Markov chains, the role of harmonic functions is played by the regular functions. A regular function f is such that $0 \leq f < \infty$ and the following condition is satisfied:

$$Qf = f,$$

where Q is the stochastic matrix of the chain [78].

For Borel right processes, h is harmonic if and only if

$$\mathbf{P}_{K^c}^- h = h, \quad m\text{-a.e.}$$

for every compact K (with the complement $K^c = \mathbf{X} \setminus K$) in an appropriate compactification of $\mathbf{X}$, where $\mathbf{P}_{K^c}^-$ is the hitting operator associated to K^c

$$\mathbf{P}_{K^c}^- h = \mathbb{E}_x \left[h(x_{T_{K^c}^-}) \right],$$
$$T_{K^c}^- = \inf \left\{ t \,|\, 0 < t < \zeta; x_{t-} \in K^c \right\}$$

(see [98], Proposition 4.4).

Theorem 7.3 (Riesz decomposition) [102] *Let* $f \in \mathcal{E}_{\mathbf{M}}$. *Then there exist a measure* μ *on* $(\mathbf{X}, \mathcal{B})$ *and a harmonic function* h *such that*

$$f = U\mu + h, \quad m\text{-a.e.} \tag{7.9}$$

Moreover, μ *is unique and* h *is unique* m-*a.e.*

From Assumption 7.1 and the characterisation of the potential $U\mu$ (Proposition 4.2 [98]), it follows that in decomposition (7.9), if there exists a compact set K such that the representing measure μ *does not 'charge'* K^c, then the function f is harmonic on K^c, i.e.,

$$\mu\left(K^c\right) = 0 \Longrightarrow f \quad \text{is harmonic on } K^c. \tag{7.10}$$

We have seen that, in an optimal stopping problem, the value function and the smallest excessive majorant of the reward function coincide. From (7.2) and the Riesz decomposition (7.9), we conclude that *the problem of finding the maximal payoff function is equivalent to the problem of finding the representing measure μ_v of v* ($U\mu_v$ is the largest measure potential dominated by v).

Roughly speaking, the continuation region C defined by (7.5), which is the region where it is not optimal to stop, is the biggest set *not charged* by the representing measure μ_v of v, i.e.,

$$\mu_v(C) = 0.$$

So, the value function v is harmonic on C (according to (7.10)). Concluding, the following result can be derived.

Proposition 7.4 *The representing measure μ_v gives the value function v, by the Riesz representation (7.9), and the support of the representation measure gives the stopping region D, defined by (7.6), i.e.,*

$$D = \text{supp}(\mu_v).$$

These considerations open a new research vista for dealing with the optimal stopping problem associated to Borel right processes. When the reward function is given by indicator functions of a target set, the representation (7.9) is quite simple and easy to deal with.

7.5.1 How the Method Works for Stochastic Hybrid Systems

We have proved that computation of the reach set probabilities reduces to the computation of the *value function* of an optimal stopping problem, with the reward function given by the indicator function of the target set involved.

The OSP methods discussed in Sect. 7.4 can be adapted to GSHS realisations if we take into consideration the special features of a GSHS (mixture of deterministic/stochastic continuous motion with random jumps), in order to obtain specific optimal stopping methods where the randomness and hybridicity of a GSHS are clearly illustrated. On the other hand, these features can be employed in order to obtain direct approximations of the réduite (value function of the OSP). We consider that for the stochastic processes that arise in the GSHS semantics, numerical computation of the reach set probabilities as value functions for certain optimal stopping problems could be supported by the following methods:

1. Based on the 'Green' kernel representation of the value function, making use of expressions of the kernel operator existing in the literature for different diffusion processes in order to obtain a good approximation of the value function;
2. prove that the réduite of the indicator function of a measurable set is the fixed point of some specific 'jump operators' [59, 60].

Stochastic Reachability for GSHSs   Let us consider the reachability problem defined for a GSHS H. Suppose that the target set A is an open set of the state space $\mathbf{X}$. Define

$$E := \mathbf{X} \backslash A.$$

Suppose the last exit time from E is finite almost surely (the process is transient), i.e.,

$$S_E = \sup\{t \geq 0 | x_t \in E\} < \infty.$$

Then the reachability problem turns into an *exit time problem*, and then computing the reach set probabilities is equivalent to the computation of a dual probability

$$\mathbb{P}_x[x_{S_E} \in E | S_E > 0]. \tag{7.11}$$

Suppose that the assumptions from Sect. 7.5 are in force.

Proposition 7.5 (Theorem 15 [101]) *For all positive bounded Borel measurable functions f on $\mathbf{X}_\Delta$, we have*

$$\mathbb{P}_x\big[(1_E f)(x_{S_E})\big| S_E > 0\big] = \int u(x, y)(1_E f)(y)\mu(dy),$$

where μ is a measure on $\mathbf{X}$.

Therefore, for f identically equal to 1, by Proposition 7.5, the probabilities (7.11) can be represented as follows:

$$\mathbb{P}_x[x_{S_E} \in E | S_E > 0] = \int_E u(x, y)\mu_E(dy), \tag{7.12}$$

where μ_E is the *equilibrium measure* of E.

Comparing representation formula (7.12) with the Riesz decomposition of the optimal value problem from Sect. 7.5 makes it obvious that for the case when the OSP reward function is the indicator function of an open set, the corresponding value function is nothing but the potential of the equilibrium measure of the complement of that open set. Then the stopping set in this case is the support of this equilibrium measure.

In order to apply this method to GSHSs, we have to perform the following computational steps:

1. computation of the kernel operator U given by (7.7), which satisfies the Poisson equation (7.8) with respect to the GSHS infinitesimal generator;

2. computation of the potential density $u(x, y)$ with respect to a reference measure (see Assumption 7.1);
3. computation of the equilibrium measure μ_E for the Riesz representation of the reach set probabilities (7.12).

In the first computational step, for PDMPs the expression of the kernel operator U (and, also, of the resolvent) can be derived based on the differential formula associated to its generator [58]. Moreover, for GSHSs, since the differential formula is also available [44], a method similar to that of [58] could be applied for solving the Poisson equation. In particular, the kernel operator, as solution of (7.8), can be approximated by the kernel operator of the embedded Markov chain. The existence of a reference measure (usually the Lebesgue measure), in the second computational step, is characterised via some specific properties of the kernel operator [228]. Carrying out the last computational step means practically (considering the previous steps) the approximation of the equilibrium measure by the analogous measure of the embedded Markov chain.

With a closer look, one can easily note that the representation of the excessive function method is related to the stochastic reachability method based on Martin capacity, developed in Sect. 6.4. Therefore, this method, even if it gives more insight in the analysis of stochastic reachability, does not provide analytic solutions supported by good computational tools. In the end, the computation of reach set probabilities should be based on Markov chain approximations.

Stochastic Reachability for PDMPs   In this section, furthermore, we investigate stochastic reachability, as an optimal stopping problem, for PDMP. The motivation for this is that, for PDMPs, characterisations of the OSP abound in the literature [59, 60, 110, 165].

The approach of [165], extended in [164], is inspired by the theory of viscosity solutions associated to first order integro-differential operators. In the above cited papers, the results are based on a 'continuity' assumption on the reset map R (associated to a PDMP). According to [63], this assumption makes the PDMP a Feller–Markov process and involves no boundary activity. So, practically, in this case, the PDMP is not a hybrid system in the traditional sense. However, the optimal control problems for Feller–Markov processes are well understood now [95] and many other results can be derived in this particular setting.

We are more interested in the approach developed in [110] and generalised in [60] and then in [59]. Mainly, in these papers the value function of the optimal stopping problem is characterised as the unique fixed point of the first jump operator. Moreover, since stochastic reachability is equivalent to an appropriate optimal stopping problem with a discontinuous reward function, the results for the OSP from [60] can be adapted for our problem in a fruitful way.

Let us recall that for the OSP studied in [60] optimal stopping is defined for a function g that is a real-valued bounded lower semianalytic function on $\mathbf{X}$ as

$$\inf_{\tau \in \Sigma} \mathbb{E}_x \big(g(x_\tau) \big).$$

Likewise, one can take the reward function g as a bounded upper semianalytic function on $\mathbf{X}$ and study the optimal stopping problem.

However, since the indicator functions are measurable functions (so, both upper/lower semianalytic), for the reachability problem we adapt the results from [60] for the value function defined using the supremum. Denote by $\mathbf{B}^*(\mathbf{X})$ the set of bounded upper semianalytic functions. Let $t^*(x)$ be the hitting time of the active boundary for the flow started in x. Define

$$\Lambda := \int_0^t \lambda\big(\varphi(s,x)\big)\,ds$$

where $0 \leq t \leq t^*(x)$.

For $x \in \mathbf{X}$ and $0 \leq t \leq t^*(x)$, for a PDMP the following standard operators can be defined [60, 64]:

(a) $J(v_1, v_2)(t, x) := \mathbb{E}_x\big[v_1\big(\varphi(t,x)\big)1_{[T_1>t]} + v_2(x_{T_1})1_{[T_1\leq t]}\big]$

$$= v_1\big(\varphi(t,x)e^{-\Lambda(t,x)}\big) + \int_0^t Rv_2\big(\varphi(s,x)\big)\lambda\big(\varphi(s,x)\big)e^{-\Lambda(s,x)}\,ds,$$

(b) $$K v_2(x) := \mathbb{E}_x\big[v_2(x_{T_1})\big],$$

(c) $$L(v_1, v_2)(x) := \left\{\inf_{0\leq t<t^*(x)} J(v_1, v_2)(t, x)\right\} \wedge K v_2(x).$$

L is called the *first jump operator*. Define the sequence of functions $(\rho_n)_{n\geq0}$ by

$$\rho_0 := g,$$

$$\rho_{n+1} := L(g, \rho_n).$$

Clearly, ρ_n is increasing and we denote by ρ its limit.

Theorem 7.6 *Let ρ_n and ρ be defined as above. Then*

(i) *the recurrence formula for ρ_n is*

$$\rho_{n+1} = L(\rho_n, \rho_n);$$

(ii) *ρ is the smallest solution of*

$$v = L(g, v), \quad v \in \mathbf{B}^*(\mathbf{X});$$

(iii) *ρ is the smallest solution of*

$$v = L(v, v), \quad v \geq g, \ v \in \mathbf{B}^*(\mathbf{X})$$

(iv) *ρ coincides with the value function defined by (7.2).*

Proof See Proposition 1, Proposition 2 and Corollary 1 from [60]. □

Proposition 7.7 *If $A \in \mathscr{B}$, the reachability function w_A is the smallest bounded upper semianalytic function of $v = L(1_A, v)$.*

Proof Take, in Theorem 7.6, $g = 1_A$ and use the characterisation of w_A as a value function for an OSP.   □

7.6 Variational Inequalities for Stochastic Reachability (I)

This section is concerned with the characterisation of reach probabilities as solutions of certain appropriate variational inequalities defined using the infinitesimal generator that is associated to a stochastic hybrid process. Since this generator is an integro-differential one, we make first an incursion into the theory of the boundary value problems corresponding to integro-differential operators.

Let **X** be a bounded open set in $\mathbb{R}^N$ with *smooth boundary*. $\mathbb{R}^N$ can be thought of as the Euclidean space where the state space of a stochastic hybrid system can be embedded.

According to [11, 164], for the existence of the viscosity solutions some assumptions are necessary. For the *Dirichlet problem* given by (7.14) and (7.15), these assumptions can be formulated as follows:

(A.1) $F \in C(\mathbb{R}^N \times \mathbb{R} \times \mathbb{R}^N \times \mathscr{S}_N \times \mathbb{R})$;
(A.2) F satisfies the local and nonlocal *degenerate ellipticity condition(s)*:

$$F(x, u, p, A, l_1) \le F(x, u, p, B, l_2) \quad \text{if } A \ge B, \ l_1 \ge l_2$$

for any $x \in \mathbb{R}^N$, $u \in \mathbb{R}$, $p \in \mathbb{R}^N$, $A, B \in \mathscr{S}^N$, $l_1, l_2 \in \mathbb{R}$;
(A.3) $R(x, \cdot)$ is a probability measure on **X** for $x \in \partial \mathbf{X}$ such that the linear operator

$$Rv(x) = \int_{\mathbf{X}} v(y) R(x, dy) \tag{7.13}$$

satisfies

$$\left| Rv(x) \right| \le C \|v\|_{L^1(\mathbf{X})}$$

for all $v \in L^1(\mathbf{X})$, where C does not depend on v;
(A.4) The function $x \longmapsto Rv(x)$ is continuous with respect to $x \in \overline{\mathbf{X}}$, uniformly for $v \in L^\infty(\mathbf{X})$.

Motivated by the expression of the generator associated to a GSHS, let us consider the linear integro-differential equations of the following form:

$$F\left(x, u, D_x u, D_x^2 u, \int_{\mathbf{X}} u(y) R(x, dy)\right) = 0, \tag{7.14}$$

where $D_x u$ denotes the space gradient, $D_x^2 u$ the matrix of second derivatives and $R(x, \cdot)$ is a probability kernel. Here, $\mathscr{S}^N$ denotes the space of symmetric $N \times N$ real-valued matrices. The applications for (7.14) are dynamic programming equations associated with the control of the right Markov processes that appear as GSHS realisations.

In the case when the state space $\mathbf{X}$ is a bounded domain of a Euclidean space, the process jumps back into $\mathbf{X}$ upon hitting the boundary, which leads to the following boundary condition to be coupled with Eq. (7.14):

$$u(x) = \int_{\mathbf{X}} u(y) R(x, dy), \quad x \in \partial\mathbf{X}. \tag{7.15}$$

For a bounded function $u : \mathbf{X} \to \mathbb{R}$, its upper/lower semicontinuous envelopes can be defined in a standard way [11, 164]. Furthermore, the definitions of the viscosity (sub/super) solutions for second order elliptic integro-differential equations are well established now in the literature [11].

Let u be a bounded function. Then:

(i) u^* is a *viscosity subsolution* of (7.14) if

$$F\left(x, u^*, D_x\phi, D_x^2\phi, \int_{\mathbf{X}} u^*(y) R(x, dy)\right) \leq 0$$

for any $\phi \in C^2(\mathbf{X})$ and any local maximum x for $u^* - \phi$.

(ii) u_* is a *viscosity supersolution* of (7.14) if

$$F\left(x, u_*, D_x\phi, D_x^2\phi, \int_{\mathbf{X}} u_*(y) R(x, dy)\right) \geq 0$$

for any $\phi \in C^2(\mathbf{X})$ and any local minimum x for $u_* - \phi$.

(iii) u is a *viscosity solution* if u is a viscosity sub- and supersolution.

A bounded function $u : \overline{\mathbf{X}} \to \mathbb{R}$ is a *viscosity* subsolution (respectively, supersolution) of the Dirichlet problem given by (7.14) and (7.15) if

(i) it is a subsolution (respectively, supersolution) of (7.14) in $\mathbf{X}$, and
(ii) for any $\phi \in C^2(\mathbf{X})$ and any local maximum (respectively, minimum) $x \in \partial\mathbf{X}$ for $u^* - \phi$ (respectively, $u_* - \phi$), it satisfies

$$\min\left\{u^*(x) - k(x), F\left(x, u^*, D_x\phi, D_x^2\phi, \int_{\mathbf{X}} u^*(y) R(x, dy)\right)\right\} \leq 0,$$

respectively,

$$\max\left\{u_*(x) - k(x), F\left(x, u_*, D_x\phi, D_x^2\phi, \int_{\mathbf{X}} u_*(y) R(x, dy)\right)\right\} \geq 0,$$

where

$$k(x) := \int_{\mathbf{X}} u(y) R(x, dy), \quad x \in \partial\mathbf{X}.$$

In general, the existence of the solutions is proved by *Perron's method*, introduced in the viscosity setting in [160]. That is, one proves that the supremum of a suitable set of subsolutions is the solution. In order to do this, one needs the help of a *comparison principle*.

In particular, for an appropriate choice of F, this Dirichlet problem becomes

$$\min\langle -\mathcal{L}u, u - g\rangle = 0 \quad \text{in } \mathbf{X}, \tag{7.16}$$

$$u(x) = \int_{\mathbf{X}} u(y) R(x, dy) \quad \text{on } \partial\mathbf{X}, \tag{7.17}$$

where $\mathcal{L}$ is the infinitesimal generator associated to a GSHS given by (4.24). Equation (7.16) with the boundary condition (7.17) is *the dynamic programming equation associated with the optimal stopping problem for a GSHS* [46]. In this case, Assumption A.2 implies that the diffusion term is nondegenerate. This is also in force in [146]. Assumption A.3 hints at the stochastic kernel R (the GSHS reset map) that should provide a bounded linear operator and Assumption A.4 involves the Feller property of the GSHS realisation [64]. For the case of Feller processes, the reward function is allowed to be semicontinuous and the value function also will be semicontinuous [12].

The main problem, in this context, is that a stochastic hybrid process is not a Feller process unless there are no active boundaries. Then these results can be applied only in some particular cases. For PDMPs, dynamic programming equations have been developed in a series of papers (see [59, 64, 87] and references therein), but, in all these papers, the reward function has some continuity properties (on the whole space or only on the trajectories of the process). Therefore, again these results cannot be applied to reachability analysis of PDMPs since the reward functions for the associated OSP do not have such continuities.

One question that arises naturally in this context is: 'How can we surmount the non-Feller characteristic of a stochastic hybrid process in order to apply dynamic programming techniques to solve the stochastic reachability problem?' We know that the non-Feller property is due to the forced (predictable) discrete transitions. Then the right approach is to 'remove' these forced transitions and to obtain switching diffusion-like processes. This research avenue was commenced in [3], but there are still many issues that are not surpassed yet. Therefore, we leave this issue as an open research problem and we will use other characterisations of the OSP that can provide more robust variational inequality characterisations.

7.7 Optimal Stopping Problem as an Obstacle Problem

This section gives an intuitive perspective on the type of variational inequalities that are going to be developed further in this chapter. The grounds on which these variational inequalities will be constructed is given by the connection between the OSP and the obstacle problem.

In control theory, the specific question of finding the optimal stopping time for a stochastic process with the payoff function $\varphi(x)$ can be characterised as an obstacle problem. The obstacle problem represents a classic motivating example in the mathematical study of variational inequalities and free boundary problems [142]. This problem consists of describing the properties of minimisers of an energy functional. A classical example of an energy functional is the Dirichlet energy, corresponding to the Laplace operator:

$$J = \int_D (\nabla u)^2 \, dx.$$

In the simple case, when the process is a Brownian motion and the process is forced to stop upon exiting the domain, the solution $u(x)$ of the obstacle problem can be characterised as the expected value of the payoff, starting the process at x, if the optimal stopping strategy is followed. The stopping criterion is simply that one should stop upon reaching the contact set. Note that the infinitesimal generator for the Brownian motion is given by the Laplace operator.

Suppose that the following data are given: (a) an open bounded domain $D \subset \mathbb{R}^n$ with smooth boundary; (b) a smooth function $f(x)$ on ∂D (the boundary of D); (c) a smooth function $\varphi(x)$ defined on all points of D such that $\varphi|_{\partial D} < f$, i.e., the restriction of φ to the boundary of D (its trace) is less than f.

Then consider the set

$$K = \left\{ u \in H^1(D) \,\middle|\, u|_{\partial D} = f \text{ and } u \geq \varphi \right\},$$

which is a closed convex subset of the Sobolev space $H^1(D)$, containing precisely those functions with the desired boundary conditions that are also above the obstacle. $H^1(D)$ is the space of square integrable functions with square integrable weak first derivatives. The solution to the obstacle problem is the function that minimises the *energy integral*

$$J(u) = \int_D |\nabla u|^2 \, dx \tag{7.18}$$

over all functions $u(x)$ belonging to K. The existence of such a minimiser is ensured by considerations of Hilbert space theory.

The obstacle problem can be reformulated as a problem in variational inequalities on Hilbert spaces. Finding the energy minimiser over the set K of suitable functions is equivalent to solving

* $u \in K$ such that

$$\int_D \nabla u \cdot \nabla (v - u) \, dx \geq 0$$

for all $v \in K$.

This is a special case of the more general form for variational inequalities on Hilbert spaces, whose solutions are functions u in some closed convex subset K of the overall space such that

$$\mathscr{E}(u, v - u) \geq h(v - u)$$

for coercive, real-valued, bounded bilinear forms $\mathscr{E}(u, v)$ and bounded linear functionals $h(v)$.

7.8 Energy Form

This section provides the formal definitions for the energy form and the concepts related to it. Let μ be a σ-finite Borel measure on $(\mathbf{X}, \mathcal{B})$. Let $L^2(\mathbf{X}, \mu)$ be the space of square integrable μ-measurable extended real-valued functions on $\mathbf{X}$, with respect to the natural inner product given by (2.31). This is a Hilbert space with respect to this inner product. First, we define the concept of (Dirichlet) energy form as follows.

Definition 7.8 We say that a map

$$\mathcal{E} : L^2(\mathbf{X}, \mu) \to \mathbb{R}^+ \cup \{\infty\}$$

is a *(Dirichlet) energy form* if the following five axioms are fulfilled:

(i) *Quadratic contraction property*: If $\mathcal{E}(f) < \infty$ and $\varphi : \mathbb{R} \to \mathbb{R}$ is an L-Lipschitz map, then

$$\mathcal{E}(\varphi \circ f) \le L^2 \mathcal{E}(f).$$

(ii) *Closedness*: If (u_n) is a Cauchy sequence in $L^2(\mathbf{X}, \mu)$ so that $\mathcal{E}(u_n) < \infty$ for all n and $\mathcal{E}(u_n - u_m) \to 0$ as $n, m \to \infty$, then $\mathcal{E}(u_n - u) \to 0$ as $n \to \infty$. Here u is the L^2-limit of the Cauchy sequence (u_n).

(iii) *Density condition*: Denoting by $D(\mathcal{E})$ the collection of all functions f in $L^2(\mathbf{X}, \mu)$ for which $\mathcal{E}(f) < \infty$, we find that $D(\mathcal{E})$ is dense in $L^2(\mathbf{X}, \mu)$ in its norm and is dense in the space $Lip_0(\mathbf{X})$ of Lipschitz functions with compact support with the supremum norm.

(iv) *Minkowski inequality*: For every pair of functions u, v in $L^2(\mathbf{X}, \mu)$:

$$\sqrt{\mathcal{E}(u + v)} \le \sqrt{\mathcal{E}(u)} + \sqrt{\mathcal{E}(v)}.$$

(v) *Parallelogram rule*: For every pair of functions u, v in $L^2(\mathbf{X}, \mu)$:

$$\mathcal{E}(u + v) + \mathcal{E}(u - v) = 2\big[\mathcal{E}(u) + \mathcal{E}(v)\big].$$

The set $D(\mathcal{E})$ is called the *Dirichlet domain corresponding to* $\mathcal{E}$. This is a Banach space with respect to the following norm:

$$\|u\|_{\mathcal{E}} := \|u\|_{L^2(\mathbf{X}, \mu)} + \sqrt{\mathcal{E}(u)}.$$

Using an energy form, one can define a bilinear form on $L^2(\mathbf{X}, \mu)$ with $D(\mathcal{E})$ dense in $L^2(\mathbf{X}, \mu)$, denoted also by $\mathcal{E}$, by the polarisation as follows:

$$\mathcal{E}(u, v) := \frac{1}{4}\big[\mathcal{E}(u + v) - \mathcal{E}(u - v)\big]. \tag{7.19}$$

This bilinear form has some remarkable properties that derive from the density condition and the quadratic contraction property. It can be shown that $\mathcal{E}$ is a *coercive closed form* on $L^2(\mathbf{X}, \mu)$, i.e., it satisfies:

(i) *Closedness axiom:* Its symmetric part

$$\widetilde{\mathscr{E}}(u, v) = \frac{1}{2}\big(\mathscr{E}(u, v) + \mathscr{E}(v, u)\big) \tag{7.20}$$

is positive definite and closed in $L^2(\mathbf{X}, \mu)$.

(ii) *Continuity axiom:* There exists a constant K such that

$$\big|\mathscr{E}_1(u, v)\big| \le K\big[\mathscr{E}_1(u, u)\big]^{1/2}\big[\mathscr{E}_1(v, v)\big]^{1/2},$$

where

$$\mathscr{E}_1(u, v) = \mathscr{E}(u, v) + \langle u, v \rangle_\mu.$$

Moreover, it can be shown that $\mathscr{E}$ satisfies the third axiom:

(iii) *Contraction condition:* For all $u \in D(\mathscr{E})$, we have

$$u^* = u^+ \wedge 1 \in D(\mathscr{E})$$

and

$$\mathscr{E}\big(u \pm u^*, u \mp u^*\big) \ge 0.$$

Remark 7.9 A coercive closed form that satisfies the contraction condition is called a (semi-) Dirichlet form.

Conversely, if a bilinear form $\mathscr{E}$ on $L^2(\mathbf{X}, \mu)$ with $D(\mathscr{E})$ dense in $L^2(\mathbf{X}, \mu)$ satisfies the axioms (i), (ii), (iii) from above, one can define, in a canonical way, the corresponding energy form:

$$\mathscr{E}(u) := \mathscr{E}(u, u), \quad u \in D(\mathscr{E}).$$

7.8.1 Energy Form Associated to a Borel Right Process

The notion of an infinitesimal generator of a Markov process is central and widely used in the theory of Markov processes to describe their properties by analytic means. As we have discussed in Chap. 6 (see also formula (2.32) in Chap. 2), a bilinear form $\mathscr{E}$ can be associated to the generator $\mathscr{L}$ of a Markov process in a natural way. The bilinear form $\mathscr{E}$ given by

$$\mathscr{E}(f, g) := \langle -\mathscr{L}f, g \rangle_\mu, \quad f \in D(\mathscr{L}), \quad g \in L^2(\mathbf{X}, \mu). \tag{7.21}$$

Form (7.21) defines a coercive closed form. For a Borel right process, the associated bilinear form, given by (7.21), might be symmetric or nonsymmetric, depending on the reversibility in time of the process (or, formally, depending on the Hermitian character of the infinitesimal generator: whether it is a self-adjoint operator or not). Due to the possible nonsymmetry of the generator, the representation formulae for the bilinear form associated to a right process might be lengthy and difficult to express and not really useful for the purposes of this approach. For the interested reader, such formulae can be found in [179]. Considering this argument, we prefer to give here only the expression of the energy form, which can be given in a unitary way in both symmetric and nonsymmetric cases.

Proposition 7.10 [179] *The energy form $\mathscr{E}$ associated to a Borel right process (x_t) can be decomposed as follows:*

$$\mathscr{E}(u) = \mathscr{E}^c(u) + \int_{\mathbf{X}} u^2 k(dx) + \int\int_{\mathbf{X} \times \mathbf{X} \setminus d} \left[u(x) - u(y)\right]^2 j(dx, dy) \quad (7.22)$$

where $\mathscr{E}^c$ corresponds to the diffusion part (associated to the local part of the infinitesimal generator), $k(dx)$ is the killing measure, which describes whether the process trajectories go the cemetery state and $j(dx, dy)$ is the jumping measure, which describes the process jumps (associated to the nonlocal part of the infinitesimal generator). These terms are uniquely determined by $\mathscr{E}$. k is a positive Radon measure and j is a positive Radon measure defined on $\mathbf{X} \times \mathbf{X} \setminus d$, where d is the diagonal of $\mathbf{X} \times \mathbf{X}$.

The energy form (7.22) may be used to derive (if it necessary) the expression of its coercive closed form. In specific problems like the obstacle problem, we may need only the expression of this energy, as in the case of Brownian motion (see (7.18)).

7.9 Variational Inequalities on Hilbert Spaces

The obstacle problem is deeply related to the study of minimal surfaces and to the *capacity* of a set in potential theory. For the example of Brownian motion discussed in Sect. 7.7, the capacity of D is

$$\text{cap}_\varphi(D) = \inf_{u \in K} J(u),$$

where J is the energy integral defined by (7.18). In a similar way, a capacity cap($\cdot$) could be introduced for a coercive closed form or for a Markov process (see Chap. 6 and [176]). We have seen elsewhere (Chap. 5) that reach set probabilities in the context of stochastic reachability can be characterised using the concept of capacity associated to a Markov process. Therefore, connections with the OSP and then the obstacle problem seem quite natural.

Let Σ be the set of stopping times and $g \in D(\mathscr{E})$. Consider the following discounted optimal stopping problem (dOSP):

$$\Phi_\alpha(x) := \sup_{\tau \in \Sigma} \mathbb{E}_x e^{-\alpha \tau} g(x_\tau), \quad \alpha > 0. \tag{7.23}$$

The aim of this section is to study certain variational inequalities associated to the coercive closed form $\mathscr{E}$ and to show that the value function Φ is the minimum solution of a variational inequality.

Set

$$\mathscr{E}_\alpha := \mathscr{E} + \alpha\langle\cdot,\cdot\rangle_\mu, \quad \alpha > 0.$$

Let us define

$$K_g := \{u \in D(\mathscr{E}) | u \geq g\}. \tag{7.24}$$

The main scope of this subsection is to solve the following *variational inequality*, i.e., to find $u_\alpha \in D(\mathscr{E})$ that satisfies

$$
\begin{aligned}
&u_\alpha \in K_g, \\
&\mathscr{E}_\alpha(u_\alpha, \phi) \geq 0, \quad \forall \phi \geq 0, \phi \in D(\mathscr{E}).
\end{aligned}
\tag{7.25}
$$

For $\varepsilon > 0$, let us consider the *penalised problem*:

$$(\alpha - \mathscr{L})(u_\alpha)_\varepsilon - \frac{1}{\varepsilon}\left(g - (u_\alpha)_\varepsilon\right)^+ = 0. \tag{7.26}$$

Lemma 7.11 *The penalised problem admits a unique solution.*

Theorem 7.12 *For any $\alpha > \alpha_0$, the variational inequality* (7.26) *admits a solution in $D(\mathscr{E})$.*

Proof We show that

$$u_{\alpha 0} = \lim_{\varepsilon \to 0}(u_\alpha)_\varepsilon$$

is solution for (7.26). Here $(u_\alpha)_\varepsilon$ is solution for (7.26). From the penalised problem, we get

$$
\begin{aligned}
\mathscr{E}_\alpha\left(g, g - (u_\alpha)_\varepsilon\right) &= \mathscr{E}_\alpha\left(g - (u_\alpha)_\varepsilon, g - (u_\alpha)_\varepsilon\right) \\
&\quad + \frac{1}{\varepsilon}\langle\left(g - (u_\alpha)_\varepsilon\right)^+, g - (u_\alpha)_\varepsilon\rangle.
\end{aligned}
$$

Note that

$$\langle\left(g - (u_\alpha)_\varepsilon\right)^+, g - (u_\alpha)_\varepsilon\rangle = \left\|\left(g - (u_\alpha)_\varepsilon\right)^+\right\|_{L^2(\mu)}$$

and that there exists a constant $\tilde{K} > 0$, such that

$$\left|\mathscr{E}_\alpha(u, v)\right| \leq \tilde{K}\sqrt{\mathscr{E}_\alpha(u, u), \mathscr{E}_\alpha(v, v)}$$

for $u, v \in D(\mathscr{E})$. Then we obtain

$$
\begin{aligned}
&\mathscr{E}_\alpha\left(g - (u_\alpha)_\varepsilon, g - (u_\alpha)_\varepsilon\right) + \frac{1}{\varepsilon}\langle\left(g - (u_\alpha)_\varepsilon\right)^+, \left(g - (u_\alpha)_\varepsilon\right)^+\rangle \\
&= \mathscr{E}_\alpha\left(g, g - (u_\alpha)_\varepsilon\right) \\
&\leq \tilde{K}\mathscr{E}_\alpha(g, g) + 1/2 E_\alpha\left(g - (u_\alpha)_\varepsilon, g - (u_\alpha)_\varepsilon\right).
\end{aligned}
$$

Now, we obtain

$$\frac{1}{2}\left[\mathscr{E}_\alpha\big(g - (u_\alpha)_\varepsilon, g - (u_\alpha)_\varepsilon\big) + \frac{1}{\varepsilon}\big\langle\big(g - (u_\alpha)_\varepsilon\big)^+, \big(g - (u_\alpha)_\varepsilon\big)^+\big\rangle\right]$$
$$\leq \widetilde{K}\mathscr{E}_\alpha(g, g), \tag{7.27}$$

and, in particular, we get

$$\big\langle\big(g - (u_\alpha)_\varepsilon\big)^+, \big(g - (u_\alpha)_\varepsilon\big)^+\big\rangle \leq 2\varepsilon\widetilde{K}\mathscr{E}_\alpha(g, g) \to 0.$$

By Fatou's lemma, we conclude that

$$(u_\alpha)_0 = \lim_{\varepsilon \to 0}(u_\alpha)_\varepsilon \geq g.$$

Next we prove that $(u_\alpha)_\varepsilon$ converges weakly to $(u_\alpha)_0$ in the Dirichlet space $(\mathscr{E}, D(\mathscr{E}))$. From (7.27), it can be shown that

$$\sup_\varepsilon \mathscr{E}_\alpha\big((u_\alpha)_\varepsilon, (u_\alpha)_\varepsilon\big) < \infty,$$

which implies that any sequence $((u_\alpha)_\varepsilon)$ contains a weakly convergent subsequence in the Dirichlet space $(\mathscr{E}, D(\mathscr{E}))$. Since

$$(u_\alpha)_\varepsilon \nearrow (u_\alpha)_0,$$

we see that $(u_\alpha)_\varepsilon$ converges weakly to $(u_\alpha)_0$ in $(\mathscr{E}, D(\mathscr{E}))$. Using the penalised problem, and observing that

$$\big(g - (u_\alpha)_\varepsilon\big)^+ = (u_\alpha)_\varepsilon \vee g - (u_\alpha)_\varepsilon,$$

and, for $\phi \geq 0$, we obtain

$$\mathscr{E}_\alpha\big((u_\alpha)_\varepsilon, \phi\big) = \frac{1}{\varepsilon}\big(\langle(u_\alpha)_\varepsilon \vee g, \phi\rangle - \langle(u_\alpha)_\varepsilon, \phi\rangle\big) \geq 0.$$

Letting $\varepsilon \to 0$, we get

$$\mathscr{E}_\alpha\big((u_\alpha)_0, \phi\big) \geq 0,$$

and the proof is over. □

Theorem 7.13 *The value function Φ_α of the dOSP (7.23) is the minimum solution for the variational inequality (7.26) and the optimal stopping time τ_α^* is the first exit time from the continuation region $\{\Phi_\alpha \leq g\}$.*

Proof The proof is similar to [186], but it has to consider the symmetric part (7.20) of $\mathscr{E}$. □

Remark 7.14 The solution Φ_α of the dOSP (7.23) belongs to the domain of $\mathscr{E}$, so it is not a strong solution in the classical sense. $D(\mathscr{E})$ can be thought of as the domain of the viscosity solutions for the variational inequality (7.26).

It remains to study the asymptotic behaviour of the solutions u_α of the variational inequality (7.26) as $\alpha \to 0$, in order to apply them to the study of the OSP with the value function given by (7.2). Such problems have been investigated analytically in several papers [94, 110, 165] for piecewise deterministic Markov processes.

Let us consider the following variational inequality, i.e., to find $u \in D(\mathscr{E})$ that satisfies

$$u \in K_g,$$
$$\mathscr{E}(u, \phi) \geq 0, \quad \forall \phi \geq 0, \ \phi \in D(\mathscr{E}). \tag{7.28}$$

Based on properties of extended Dirichlet spaces [93], we can write:

Theorem 7.15 *The value function Φ of the OSP (7.2) is the minimum solution for the variational inequality (7.28) and the optimal stopping time τ^* is the first exit time from the continuation region $\{\Phi \leq g\}$.*

Using the arsenal of energy forms theory, it is possible to show that a solution u_α of the variational inequality (7.26) weakly converges to a solution of (7.28). Therefore, we can conclude that Φ_α converges quasi-everywhere to Φ as α tends to zero.

7.10 Variational Inequalities for Stochastic Reachability (II)

In this section, we study the reachability problem in terms of optimal stopping for different particular cases of stochastic hybrid systems. Stochastic hybrid processes are Markov processes whose infinitesimal generator can be decomposed in a local part (corresponding to the continuous evolution described by certain diffusion processes) and a nonlocal part (corresponding to the discrete transitions described by a jump process). For the Markov process **M** that represents the realisation of a GSHS, H, the energy form can be given as

$$\mathscr{E}(u) = \mathscr{E}^c(u) + \lambda(x) \iint_{X \times X \backslash d} \left[u(x) - u(y) \right]^2 j(dx, dy) \tag{7.29}$$

where $\mathscr{E}^c$ corresponds to the diffusion part and j corresponds to the jumping part, where the $D(\mathscr{E})$ is included in the domain of the generator $\mathscr{L}$. We will reveal the expressions of $\mathscr{E}^c$ and j after we will treat the particular cases of diffusions and jump processes in the following subsections.

The variational inequality (7.28) also characterises stochastic reachability. Let $A \in \mathscr{B}(\mathbf{X})$ and g be its indicator function. To avoid the complications related to the boundary condition (7.15) of the domain of the generator, we suppose that either one of the following assumptions is true:

(B.1) If $A \cap \partial\mathbf{X} = \emptyset$ then the hybrid process never jumps in A, i.e., the reset map R satisfies the condition:

$$R(x, A) = 0, \quad \forall x \in \partial\mathbf{X}.$$

(B.2) If $A \cap \partial \mathbf{X} \neq \emptyset$ then the hybrid process has a jump/switching from the boundary of A into A, i.e., the reset map R satisfies the condition:

$$R(x, A) = 1, \quad \forall x \in A \cap \partial \mathbf{X}.$$

Then we can derive the following result.

Theorem 7.16 *If (B.1) or (B.2) is true, then the reachability function w_A is the minimum of the following variational inequality*:

$$u \in K_g;$$
$$\mathscr{E}(u + \phi) \geq \mathscr{E}(u - \phi), \quad \forall \phi \geq 0, \ \phi \in D(\mathscr{E}),$$

where K_g is defined by (7.24) and $\mathscr{E}$ is the energy form associated to a stochastic hybrid process.

Proof If, in Theorem 7.15, we take g equal to the indicator function of the target set and use formula (7.19), we easily obtain the desired characterisation of the reachability function. □

Remark 7.17 Note that this indicator function is not in $D(\mathscr{E})$; therefore, we need to consider its excessive regularisation that belongs to $D(\mathscr{E})$. This regularisation is the smallest excessive function that dominates it, which is different from this indicator function only on a negligible set [46].

Practically, computing reach set probabilities means to solve an obstacle problem with an irregular obstacle, which is not continuous and does not belong to a Sobolev space. Despite this irregular behaviour, we are still able to include the solutions in a quite nice set of functions from the domain of the energy form. The 'smoothness' of the solutions can be described in terms of the regularity of the energy forms [176].

7.10.1 Reachability for Diffusion Processes

Consider a finite-dimensional diffusion process with the state space $\mathbb{R}^d$, whose coefficients are given by a vector field $b(z) = (b_1(z), \dots, b_d(z))$ and a matrix $(\sigma_{i,j}(z))_{1 \leq i, j \leq d}$. The reference measure will be the Lebesgue measure and the closed coercive form is defined as follows:

$$\mathscr{E}(u, v) = \sum_{i.j=1}^{d} \int_{\mathbb{R}^d} \sigma_{i,j}(x) \frac{\partial u}{\partial x_i} \frac{\partial v}{\partial x_j} \, dx$$

$$+ \sum_{i=1}^{d} \int_{\mathbb{R}^d} b_i(x) \frac{\partial u}{\partial x_i} v(x) \, dx, \qquad (7.30)$$

$$D(\mathscr{E}) = \left\{ u \,\middle|\, u \in L^2(\mathbb{R}^d), \nabla u \in L^2(\mathbb{R}^d) \right\}.$$

Theorem 7.15 can be applied to obtain a characterisation of the value function associated to the optimal stopping of the diffusion process associated with this bilinear form.

For diffusion processes, the OSP and the obstacle problem are well studied. The interested reader may compare our variational inequalities with those already existing in the literature. For a thorough exposition of OSPs and free boundary problems, see the recent textbook of Peskir and Shiryayev [190].

7.10.2 Reachability for Jump Processes

Fix ρ, a metric on $\mathbf{X}$ compatible with the given topology. Any jump process is described using a stochastic kernel. Suppose we have given a stochastic kernel R on $\mathbf{X} \times \mathscr{B}(\mathbf{X})$ satisfying the following two conditions:

(a) For any $\varepsilon > 0$, $R(x, \mathbf{X} \backslash U_\varepsilon(x))$ is, as a function of $x \in \mathbf{X}$, locally integrable with respect to a measure m. Here, $U_\varepsilon(x)$ is the ε-neighbourhood of x.

(b)

$$\int_{\mathbf{X}} u(x)(Rv)(x)\mu(dx) = \int_{\mathbf{X}} v(x)(Ru)(x)\mu(dx) \leq \infty$$

for $u, v \in \mathscr{B}^b(\mathbf{X})$.

We have assumed that R satisfies the symmetry axiom (b). This assumption can be fulfilled in the practical application considering its symmetrised form. Then R determines a positive Radon measure $j(dx, dy)$ on $\mathbf{X} \times \mathbf{X} \backslash d$ by

$$\int_{\mathbf{X} \times \mathbf{X} \backslash d} f(x, y) j(dx, dy) = \int_{\mathbf{X}} \left[\int_{\mathbf{X}} f(x, y) R(x, dy) \right] m(dy),$$

where $f \in C_0(\mathbf{X} \times \mathbf{X} \backslash d)$ (set of continuous functions with compact support on $\mathbf{X} \times \mathbf{X} \backslash d$). The symmetry of R ensures the symmetry of j. Then the coercive form associated to the jump process is

$$\mathscr{E}(u, v) = \int \int_{\mathbf{X} \times \mathbf{X} \backslash d} \left[u(x) - u(y) \right] v(x) - v(y) j(dx, dy), \tag{7.31}$$

$$D(\mathscr{E}) = \left\{ u \in L^2(\mathbf{X}, \mu), \mathscr{E}(u) < \infty \right\},$$

where its energy form $\mathscr{E}$ is given by

$$\mathscr{E}(u) = \int \int_{\mathbf{X} \times \mathbf{X} \backslash d} \left[u(x) - u(y) \right]^2 j(dx, dy). \tag{7.32}$$

It is easy to see that the energy form (7.32) may be defined in a similar way for a stochastic kernel R, which is not symmetric.

From (7.30) and (7.31), it is easy to identify all the terms that appear in the expression of the energy form (7.29) for a stochastic hybrid process. Note that the Dirichlet domain of this form should be defined carefully as the set of all functions belonging to $D(\mathscr{L})$ (domain of the infinitesimal generator) for which this energy is finite.

7.11 Hamilton–Jacobi–Bellman Equations

This section addresses stochastic reachability and safety problems for GSHSs based on the characterisation of reachability probabilities as viscosity solutions of a system of coupled Hamilton–Jacobi–Bellman (HJB) equations. Based on this formulation, we can derive a computational method based on discrete approximations for solving reachability analysis problems for GSHSs. The material presented in this section is based on [146], but notations and proofs have been carefully analysed and improved. This section provides a new characterisation of the reach probability function as a value function of a dynamic programming problem. Moreover, this function can be characterised as a fixed point of a recursive operator defined with respect to the stopping times that represent the jumping times of the underlying stochastic hybrid process. Under the nondegeneracy assumption for the component diffusions that describe the continuous dynamics in modes, the value function is bounded and continuous in each mode of the given GSHS. Hence, we can derive that the value function for the reachability problem of the GSHS is locally equal to the value function for the exit problem of a standard stochastic diffusion, but the boundary conditions depend on the value function itself. Based on this formulation, the reachability function is represented as a viscosity solution of a system of coupled HJB equations.

Let

$$H = \big((Q, d, \mathcal{X}), b, \sigma, \text{Init}, \lambda, R\big)$$

be a standard GSHS, with its Markov realisation **M**.

Assumption 7.2 (Nondegeneracy) [146] The boundaries ∂X^q are assumed to be sufficiently smooth and the trajectories of the system satisfy a nontangency condition with respect to the boundaries. A sufficient condition for the nontangency assumption is that the diffusion term is nondegenerate, i.e.,

$$a(q, x) = \sigma(q, x)\sigma^{\mathsf{T}}(q, x), \quad q \in Q$$

is positive definite.

This standard assumption is used to ensure the continuity of the viscosity solution close to the boundaries [90]. Discussions of other kinds of condition that ensure such continuity can be found in [146].

Assumption 7.3 (Boundedness) [146] The set of locations Q is finite and that X^q is bounded for every q.

Usually, most practical systems have finitely many modes and boundedness constraints on the continuous state. Even if the state space is unbounded, in many cases we would like to approximate it for numerical computation purposes. This assumption is used for approximating the underlying hybrid system by a finite state space Markov chain so we may apply numerical methods based on dynamic programming.

Suppose that $A \in \mathscr{B}(\mathbf{X})$ is a target set and $B \in \mathscr{B}(\mathbf{X})$ is an 'unsafe' set in the hybrid state space $\mathbf{X}$. In general, for a GSHS, these sets may be described as unions of target and unsafe sets, respectively, for multiple modes. For example, a target set may be represented as a subset of the continuous state space, which requires that we consider a target subset for every discrete state. Then we have the following representations:

$$A := \bigcup_{q \in Q} \{q\} \times A_q,$$

$$B := \bigcup_{q \in Q} \{q\} \times B_q.$$

We suppose that, for each $q \in Q$, the sets A_q and B_q are proper open subsets of the modes X^q, i.e., their boundaries do not intersect the boundaries of X^q. We define

$$E_q := X^q \backslash (\overline{A}_q \cup \overline{B}_q);$$

$$E := \bigcup_{q \in Q} \{q\} \times E_q.$$

The initial state (which, in general, is governed by a probability distribution) must lie outside sets A and B. The stochastic kernel $R(\cdot, \Gamma)$ is assumed to be defined so that the system cannot jump directly to B or A, i.e.,

$$R(\cdot, A) = 0; \qquad R(\cdot, B) = 0.$$

Let us denote by T_∂, the first hitting time of $\partial A_q \cup \partial B_q$. Suppose that $\mathbf{x} \in E$ is an initial condition, then we define the reachability function as follows:

$$V(\mathbf{x}) := \begin{cases} \mathbb{P}_{\mathbf{x}}(x_{T_\partial^-} \in \partial A), & \mathbf{x} \in E, \\ 1, & \mathbf{x} \in \partial A, \\ 0, & \mathbf{x} \in \partial B. \end{cases} \qquad (7.33)$$

The function $V(\mathbf{x})$ can be interpreted as the state-constrained reach probability that a trajectory starting at x will reach set A whilst avoiding set B.

If the trajectory hits the boundary of either the unsafe or the target set, it is assumed that the execution of the GSHS terminates. The killing of the stochastic process is formalised by slightly modifying the GSHS model. This can be done by adding a cemetery point Δ. The transition to the cemetery point Δ can be captured by extending the reset kernel R, as follows:

$$R(\mathbf{x}, \Delta) := \begin{cases} 1, & \mathbf{x} \in \partial A \cup \partial B, \\ 0, & \text{otherwise.} \end{cases}$$

The reachability function V is extended by defining

$$V(\Delta) = 0,$$

which agrees with the probabilistic interpretation of V. By abuse of notation, killed process will be denoted also by $\mathbf{x}_t$. Then we consider a continuous function $c : \overline{\mathbf{X}} \to \mathbb{R}_+$ that satisfies the boundary conditions

$$c(\mathbf{x}) = \begin{cases} 1, & \mathbf{x} \in \partial A, \\ 0, & \mathbf{x} \in \partial B \cup \partial \mathbf{X}. \end{cases}$$

The function c should be thought of as a smooth function that goes from 1 to 0, when $\mathbf{x}$ runs between A and B. Let $p^*(t)$ be the counting the number of jumps from the boundary of the process $(\mathbf{x}_t)$ up to time t, i.e.,

$$p^*(t) := \sum_{k=1}^{\infty} I_{(t \geq T_k)} I_{(x_{T_k^-} \in \partial \mathbf{X})}.$$

Then

$$V(\mathbf{x}) = \mathbb{E}_{\mathbf{x}} \int_0^{\infty} c(\mathbf{x}_{t^-}) \, dp^*(t).$$

With this representation in hand, we show that V can be characterised as a viscosity solution of a system of coupled HJB equations. In fact, this representation can be connected with the representation of the hitting time probabilities as solutions for boundary value problems corresponding to each infinitesimal generator associated to the component diffusions of a GSHS. (In each mode X^q, the restriction of c to X^q is the solution of the respective boundary value problem.)

Based on a dynamic programming argument, one can derive a representation of the reach probability function that resembles a discount cost criterion with a target set. The results available for standard diffusion processes can be extended to show that this function is characterised as a viscosity solution of a system of HJB equations.

Suppose that $\mathscr{G} : \mathscr{B}_+^b(\mathbf{X}) \to \mathscr{B}_+^b(\mathbf{X})$ is an operator defined by

$$\mathscr{G}g(\mathbf{x}) := \mathbb{E}_{\mathbf{x}}\big[c(\mathbf{x}_{T_1^-})I_{(\mathbf{x}_{T_1^-} \in \partial \mathbf{X})} + g(\mathbf{x}_{T_1})\big],$$

where T_1 is the first jumping time. The recursion required by the dynamic programming is defined with respect to the stopping times T_i that represent the jumping times.

Lemma 7.18 [146]

$$\mathscr{G}^n g(\mathbf{x}) := \mathbb{E}_{\mathbf{x}}\bigg[\int_0^{T_n} c(\mathbf{x}_{t^-}) \, dp^*(t) + g(\mathbf{x}_{T_n})\bigg].$$

Theorem 7.19 [146] *The reachability function V is a fixed point of operator $\mathscr{G}$.*

Let $t^*(\mathbf{x})$ be the hitting time of the active boundary for the execution started in $\mathbf{x}$. Define the stochastic process

$$\Lambda_t := \exp\bigg(-\int_0^t \lambda(\mathbf{x}_s) \, ds\bigg),$$

where $0 \leq t \leq t^*(\mathbf{x})$. Then the following notation is necessary:

$$LV(\mathbf{x}) := \lambda(\mathbf{x}) \int_E V(\mathbf{y}) R(\mathbf{x}, d\mathbf{y}),$$

$$JV(\mathbf{x}) := c(\mathbf{x}) + \int_E V(\mathbf{y}) R(\mathbf{x}, d\mathbf{y}),$$

where V is the reachability function given by (7.33).

Theorem 7.20 [146] *The reachability function V given by (7.33) admits the following representation:*

$$V(\mathbf{x}) = \mathbb{E}_{\mathbf{x}} \left[\left(\int_0^{T_1} \Lambda_t LV(\mathbf{x}_{t-}) \, dt + \Lambda_{T_1} JV(\mathbf{x}_{T_1}) \right) I_{(\mathbf{x}_{T_1^-} \in \partial \mathbf{X})} \right]$$

for all $\mathbf{x} \in E$.

Theorem 7.21 [146] *V is bounded and continuous in $\mathbf{x}$ on $\overline{E}$.*

The proof of this result is a consequence of the fact that V is locally a solution of a Dirichlet boundary value problem associated with the infinitesimal generator of the local diffusion. This result is not so surprising, since we consider only the restriction to $\overline{E}$ of the reachability function V. A direct proof is provided in [146].

Assumption 7.4 Assume that b and σ are continuously differentiable with respect to x in E_q for each q and, for suitable C_1 and C_2, satisfy

$$|b(q, x)| \le C_1, \qquad |\sigma(q, x)| \le C_1$$

and

$$|b(q, 0)| + |\sigma(q, x)| \le C_2.$$

Theorem 7.22 (HJB equations) [146] *Under the standard hypotheses that ensure the existence and uniqueness of realisation $\mathbf{M}$ of a GSHS H and Assumption 7.4, V is the viscosity solution of the following Hamilton–Jacobi–Bellman equation:*

$$\mathscr{H}\big((q, x), D_x V(q, x), D_x^2 V(q, x)\big) = 0, \quad x \in E_q, \ q \in Q$$

with boundary conditions

$$V(q, x) = JV(q, x), \quad x \in \partial E_q, \ q \in Q$$

where the Hamiltonian operator is given by

$$\mathscr{H}\big((q, x), D_x \varphi, D_x^2 \varphi\big) = b(q, x) D_x \varphi + \frac{1}{2} \mathrm{tr}\big(a(q, x) D_x^2 \varphi\big) + L\varphi(q, x)$$

$$- \lambda(q, x)\varphi(q, x)$$

for any test function $\varphi \in C^2(X^q)$.

In [146], the proof of this theorem is a consequence of similar results for standard diffusions developed in [90]. Furthermore, the continuity of V stated by Theorem 7.21 ensures that

$$V(q,x) = \begin{cases} \mathscr{G}V(q,x) & x \in E_q, \\ JV(q,x), & x \in \partial E_q \end{cases}$$

is the unique continuous solution of the HJB equation corresponding to the mode X^q.

This method has been applied for a stochastic version of the thermostat (for n rooms) benchmark model and the results are presented in [146]. This particular example models the dynamics of a temperature in a building with three rooms. The temperature x_i in a room ($i = 1, 2, 3$) depends on the temperature of the adjacent rooms, the outside temperature u and the location of the heater (if it is placed in the room and turned on). The SDE describing the continuous dynamics of the temperature is

$$dx = (Ax + Bu + Cq)\,dt + \Sigma\,dw$$

where

$$A = \begin{bmatrix} -0.9 & 0.5 & 0 \\ 0.5 & -1.3 & 0.5 \\ 0 & 0.5 & -0.9 \end{bmatrix} \quad \text{and} \quad B = \begin{bmatrix} 0.4 \\ 0.3 \\ 0.4 \end{bmatrix}$$

$C = \mathrm{diag}(6, 7, 8)$, $u = 4$, $\Sigma = \mathrm{diag}(0.1)$, q is a vector of 0s and 1s, representing the position and state (off/on) of the heaters and $w(t)$ is an $\mathbb{R}^3$-valued Wiener process.

The discrete states of the system describe the position and condition of the heaters in their respective rooms. When a heater is in a room and switched on, then a 1 is placed in the corresponding position of that room. If the heater is not in the room or it is in the room but switched off, then a 0 is placed in the corresponding position of that room. Therefore, for the case of three rooms, the model has 12 modes. The discrete transitions are denoted by arcs between nodes and are defined using a control policy for moving the heater. This control policy is embedded in the discrete mode invariants. In [146], the following control policy for rooms i and j has been considered. If a heater situated in the room i is off, it is turned on if $x_i \leq 19$ and a heater that is on is turned off if $x_i \geq 20$. A heater is moved from room j to an adjacent room i if the following conditions hold: (1) room i is without a heater; (2) room j currently has a heater; (3) $x_i \leq 17$; and (4) $x_j - x_i \geq 1$.

The safe set is described by $x_i \in (10, 20)$, $i = 1, 2, 3$. A discretisation of the continuous space is required. The approximation parameter $h = 0.25$. Then with 12 discrete modes, the approximating process will have $12 \times 42^3 = 889056$ (including also the boundary of the safe set). The room heater model evolves in a 3-D continuous space. A threshold for the acceptable probability is specified to be (0.1). The safe set for each mode is the set of states that have a probability below the threshold to reach the unsafe set. The iterative algorithm has been executed in approximately 49 min on a 3.0-GHz desktop computer [146].

7.12 Some Remarks

This chapter summarised some remarkable analytic approaches on stochastic reachability. Mainly, for the characterisation of reach probabilities, variational inequalities on Hilbert spaces were derived. These are constructed on the ground of functional analysis techniques that are used in practice for studying Markov processes. Variational inequalities based on energy forms, Dirichlet boundary problems and Hamilton–Jacobi–Bellman equations provide solutions for reachability probability estimations. The basic attack of all these methods depends on original ways to handle the infinitesimal generator of a stochastic hybrid process. It is known that energy forms and the Hamilton operators are closely related. Dynamic programming equations have equivalent formulations in both approaches. Moreover, further developments can be investigated using the transition semigroup and the operator resolvent.

Chapter 8
Statistical Methods to Stochastic Reachability

8.1 Overview

It was proven (Chap. 7) that for reach set probabilities it is hard to find analytical formulae. At least, one might obtain upper bounds of these, but again these bounds cannot be easily computed. So it seems we must set up a statistical framework, which will allow us to find suitable algorithms to compute these probabilities.

In this chapter, we characterise the stochastic reachability in terms of Bayesian statistics. This will permit us to employ statistical inference rules in order to compute the reach set probabilities that appear in the formulation of stochastic reachability analysis. To achieve this task, the stochastic reachability concept will be expressed in terms of imprecise probabilities. Here, imprecision in probability assessments is modelled through convex sets of probability measures. This is possible since we consider only stochastic hybrid systems whose realisations are Markov processes with nice properties. Then the reach set probabilities define a Choquet capacity. The latter concept is widely used in decision theory [68, 209] and robust Bayesian inference [120], and it is closely related to other concepts modelling different kinds of probability set such as: lower probabilities [88], belief functions [68], lower envelopes [120], lower expectations and lower previsions [227].

Objectives   The objectives of this chapter can be summarised as follows:

1. To give a quick introduction in the area of imprecise probabilities.
2. To set up a Bayesian framework for a particular class of imprecise probabilities, namely Choquet capacities.
3. To characterise the stochastic reachability problem in terms of Choquet capacities.
4. To propose a way to calculate reach set probabilities using different algorithms used in statistical inference (for the computation of conditional upper expectations).

This chapter is only a beginning of the research in this area. The theory of imprecise probabilities is well developed and offers a palette of original methods that allow us to explore and understand these probabilities.

L.M. Bujorianu, *Stochastic Reachability Analysis of Hybrid Systems*,
Communications and Control Engineering,
DOI 10.1007/978-1-4471-2795-6_8, © Springer-Verlag London Limited 2012

8.2 Imprecise Probabilities

Decision theory starts with the states, acts and utilities that have to be specified by the acting agent. Making a decision entails choosing a series of steps to follow. Bayesian theory has been very successful as a prescription for what a rational agent should do [206]. The Bayesian framework essentially says that:

- Given the states of nature θ_i, there is a single probability distribution $p(\theta)$ that summarises the beliefs of the agent from which θ_i is obtained.
- An act with high expected utility is preferred to an act with lower expected utility.

The Bayesian framework is built on a number of axioms that are supposed to apply to decision making. To leverage this framework such that we can deal with nonlinear expectation, it is necessary to start with a similar, but more general set of axioms. Then the next step is to generate a convex set of probability distributions, called a *credal set* [166]. In this context, Bayesian theory describes a particular scenario, in which we assume that the agent always has a single distribution (the convex set of distributions has a single member).

The theory of sets of probabilities advocates that a rational agent chooses an act based on expected loss considerations. An *expected loss* is defined by two entities:

1. A *loss function*, which translates the preferences of the agent. Some people use the term 'utility' for the reverse of loss, i.e., loss with a minus sign. But loss and utility are essentially the same notion.
2. A set of probability distributions called a *credal set*. Usually a credal set is assumed to be a convex set of probability distributions. A credal set conveys the beliefs of an agent about the possible states of the world.

Usually, a theory of probability set is a normative theory of decision making (i.e., it is concerned with identifying the best decision to take). So the purpose is to explain how an agent should make decisions. The agent starts with a prior credal set, uses a likelihood credal set and reaches a posterior credal set. Then the agent picks an option that minimises expected loss.

A number of theories of inference advocate closed convex sets of probability measures as an accurate representation for imprecise belief [159, 166]. Several other theories employ special types of convex sets of probability measures, for example the theory of lower probability [88] and of inner/outer measures [218]. Others are based on Choquet capacities, monotone (convex) capacities, 2-monotone capacities and infinitely monotone capacities (belief functions) [120]. The theory of coherent lower previsions (lower expectations) put forward by Walley is an example of a complete theory of inference that can be viewed as a theory of sets of probability measures [227]. There are also theories of inference that add imprecision in utility judgements to the modelling process [210]. Theories are normative because they only offer some sensible guidelines; they are not meant to be a description of how real agents work.

In the following we present Bayes' theorem for capacities. This theorem will then be applied for statistical reasoning regarding reach set probabilities.

8.2.1 Choquet Integral

Some introductory definitions on capacities were given in Chap. 5, Sect. 5.5.5. To state Bayes' theorem for capacity, we need first to introduce the (nonlinear) expectation with respect to a capacity. Therefore, we need a concept of integral with respect to a capacity.

Suppose that $c : \mathscr{A} \to [0, 1]$ is a capacity on $(\mathbf{X}, \mathscr{A})$. Since we allow the possibility that a capacity c not be additive, we cannot use the integral in the Lebesgue sense to integrate with respect to c. The notion of integral we will use is due originally to Choquet [51], and it was independently rediscovered and extended by Schmeidler [209].

If $f : \mathbf{X} \to \mathbb{R}$ is bounded an $\mathscr{A}$-measurable function and c is any capacity on $\mathbf{X}$ we define the *Choquet integral* of f *with respect to* c to be the number

$$c(f) := \int_{\mathbf{X}} f(x) \, dc(x) = \overline{c}(f) + \underline{c}(f),$$

where $\overline{c}(f)$ is the *upper integral*

$$\overline{c}(f) := \int_0^\infty c\big(\{x \in \mathbf{X} | f(x) \geq \alpha\}\big) \, d\alpha \tag{8.1}$$

and $\underline{c}(f)$ is the *lower integral*

$$\underline{c}(f) := \int_{-\infty}^0 \big[c\big(\{x \in \mathbf{X} | f(x) \geq \alpha\}\big) - 1 \big] d\alpha. \tag{8.2}$$

The integrals (8.1) and (8.2) are taken in the sense of Riemann. Then the integral $c(f)$ can be thought of as an expected utility without additivity [209].

8.2.2 Bayes' Theorem for Capacities

Let $(\mathbf{X}, \mathscr{B})$ be a Polish space, i.e., the topology for $\mathbf{X}$ is complete, separable and metrisable, and $\mathscr{B}$ is the Borel σ-algebra of $\mathbf{X}$. As usual, denote by $\mathscr{B}^b(\mathbf{X})$ the set of all bounded, nonnegative, real-valued measurable functions defined on $\mathbf{X}$.

Let $(\mathbb{P}_x | x \in \mathbf{X})$ be a set of probability measures on a sample space $(\Omega, \mathscr{F})$. Assume that each $\mathbb{P}_x$ has a density $p(x|\omega)$ with respect to some σ-finite measure and let

$$L(x) = p(x|\omega)$$

be the likelihood function for x having observed $\omega \in \Omega$. We assume that $L \in \mathscr{B}^b(\mathbf{X})$.

Let $\mathscr{P}$ be a nonempty set of prior probabilities on $\mathscr{B}$ and define the upper and lower prior probability functions by

$$\overline{\mu}(A) = \sup_{\mu \in \mathcal{P}} \mu(A), \tag{8.3}$$

$$\underline{\mu}(A) = \inf_{\mu \in \mathcal{P}} \mu(A). \tag{8.4}$$

Clearly,

$$\overline{\mu}(A) = 1 - \underline{\mu}(\mathbf{X} \backslash A),$$

then it is enough to study only $\underline{\mu}$.

Suppose that $\mathcal{P}$ is convex. Then $\overline{\mu}$ can be thought of as the capacity c generated by the family $\mathcal{P}$ (the upper envelope of $\mathcal{P}$). For each $f \in \mathcal{B}^b(\mathbf{X})$, define the *upper expectation* of f by

$$\overline{\mathbb{E}}(f) = \sup_{\mu \in \mathcal{P}} \mu(f), \tag{8.5}$$

where

$$\mu(f) = \int f(x)\mu(dx).$$

Analogously, the *lower expectation* $\underline{\mathbb{E}}(f)$ can be defined.

In a similar way, one can define the *upper Choquet integral* of f defined by

$$c_u(f) := \int_0^\infty \overline{\mu}\big(\{x \in \mathbf{X} \,|\, f(x) \geq \alpha\}\big)\, d\alpha$$

and the *lower Choquet integral* of f by

$$c_l(f) := \int_0^\infty \underline{\mu}\big(\{x \in \mathbf{X} \,|\, f(x) \geq \alpha\}\big)\, d\alpha.$$

We say that $\mathcal{P}$ *is closed with respect to majorisation*, or is *m-closed*, if

$$\mu \leq \overline{\mu} \Rightarrow \mu \in \mathcal{P}.$$

For a prior probability $\mu \in \mathcal{P}$, for which $\mu(L) > 0$, the posterior probability of a subset A, after observing ω and applying Bayes' theorem, may be expressed as

$$\mu(A|\omega) = \frac{\mu(L_A)}{\mu(L_A) + \mu(L_{A^c})},$$

where

$$L_A(x) := L(x) I_A(x)$$

and I_A is the indicator function of A.

We denote

$$\mathcal{P}(\cdot|\omega) = \big\{\mu(\cdot|\omega)\,\big|\,\mu \in \mathcal{P} \text{ with } \mu(L) > 0\big\}.$$

Next we give Bayes' theorem for Choquet capacities as a particular case of the main result of [225].

Theorem 8.1 *Let $\mathscr{P}$ be a nonempty m-closed set of prior probabilities on $\mathscr{B}$ and let $\mathscr{P}(\cdot|\omega)$ be the corresponding class of posterior probabilities. If $\mathscr{P}$ generates a submodular Choquet capacity c, then for each $A \in \mathscr{B}$, the capacity generated as the upper envelope of $\mathscr{P}(\cdot|\omega)$ can be computed as follows:*

$$
\begin{aligned}
c(A|\omega) &= \frac{\overline{\mathbb{E}}(L_A)}{\overline{\mathbb{E}}(L_A) + \underline{\mathbb{E}}(L_{A^c})} \\
&= \frac{c_u(L_A)}{c_u(L_A) + c_l(L_{A^c})}
\end{aligned}
$$

when the ratios are well defined.

Remark 8.2 If in Theorem 8.1, the capacity c is not a Choquet capacity, then both equal signs should be replaced by $\leq$. Then, in this case, Bayes' theorem for capacities gives upper bounds of the conditional capacity $c(\cdot|\omega)$.

8.3 Capacities and Stochastic Reachability

This section is the core of the chapter. After we point out the connections between the reach set probabilities and capacities, we apply Bayes' theorem for stochastic reachability analysis.

8.3.1 Capacities Associated to GSHS Realisations

Hypothesis   We assume without loss of generality that the realisation of stochastic hybrid system H,

$$
\mathbf{M} = (\Omega, \mathscr{F}, \mathscr{F}_t, x_t, \mathbb{P}_x)
$$

is a Borel right Markov process with the state space $(\mathbf{X}, \mathscr{B})$.

This is the case with most models for stochastic hybrid systems developed in the literature. We studied GSHS, which is the most general model of stochastic hybrid systems, in Chap. 4. As usual, we suppose that $\mathbf{M}$ is *transient*, and $\mathbf{X}$ is equipped with a σ-finite measure m.

Canonical Representation of the Underlying Probability Space   Let Δ be the cemetery point for $\mathbf{X}$, which is an adjoined point to $\mathbf{X}$:

$$
\mathbf{X}_\Delta = \mathbf{X} \cup \{\Delta\}.
$$

Let $\zeta(\omega)$ be the lifetime, or 'termination time', when the process $\mathbf{M}$ escapes to and is trapped at Δ.

One can take the canonical representation of the sample space Ω for $\mathbf{M}$ as the set of all its paths. Then the trajectories of the process $\mathbf{M}$ can be thought of as elements in the sample probability space. This means each $\omega \in \Omega$ is a process evolution, i.e., $\omega = (\omega_t)_{t \geq 0}$. The process (x_t) can be viewed as the signal process and (ω_t) as the observable process.

The Global Probability Measure $\mathbb{P}$ Because of the transience condition, measure m is purely excessive [97]:

$$\lim_{t \to \infty} \big(m \langle P_t \rangle \big)(A) = 0$$

for all $A \in \mathscr{B}$ with $m(A) < \infty$, where

$$\big(m \langle P_t \rangle \big)(A) := \int p_t(x, A) m(dx)$$

and

$$p_t(x, A) = P_t(I_A)(x) = \mathbb{P}_x(x_t \in A). \tag{8.6}$$

Consequently, there is a unique entrance law $(\mu_t)_{t>0}$ (a family of σ-finite measures on $(\mathbf{X}, \mathscr{B})$ with

$$\mu_t \langle P_s \rangle = \mu_{t+s}$$

for all $t, s > 0$) such that

$$m(A) = \int_0^\infty \mu_t(A) \, dt, \quad \forall A \in \mathscr{B}.$$

See, for example [97], for more details. Then there is a σ-finite measure $\mathbb{P}$ on $(\Omega, \mathscr{F}_t^0, \mathscr{F}^0)$ under which the coordinate process $(x_t)_{t>0}$ is Markovian with transition semigroup $(P_t)_{t \geq 0}$ and one-dimensional distributions

$$\mathbb{P}(x_t \in A) = \mu_t(A), \quad \forall A \in \mathscr{B}, \, t > 0.$$

The Construction of the Capacity   We learned that the *capacity associated to* $\mathbf{M}$ is

$$\mathrm{Cap}_{\mathbf{M}}(B) = \mathbb{P}(T_B < \infty), \tag{8.7}$$

for all $B \in \mathscr{B}$ (with T_B the first hitting time of B). The initial definition of this notion gives the capacity $\mathrm{Cap}_{\mathbf{M}}$ as an upper envelope of a nonempty class of probability measures on $\mathscr{B}$. Moreover, in Chap. 5 other properties of $\mathrm{Cap}_{\mathbf{M}}$ were recorded (*monotone increasing, submodular, countably subadditive*).

Then its conjugate $\mathrm{Cap}_{\mathbf{M}}^*$ [209], defined by

$$\mathrm{Cap}_{\mathbf{M}}^*(B) := 1 - \mathrm{Cap}_{\mathbf{M}}\big(B^c\big)$$

is a *belief function* in sense of [211] (here, $B^c = \mathbf{X} - B$). Beliefs about the evolution process (ω_t) conform to a time-homogeneous Markov structure. In standard models, this would involve a stochastic kernel giving conditional probabilities. We assume that beliefs conditional on ω_t are too vague to be represented by a probability measure and are represented instead by a family of probability measures whose lower envelope is $\mathrm{Cap}_{\mathbf{M}}^*$.

Let us consider $\mathscr{P}_{\mathbf{M}}$, the family of all probability measures on $(\mathbf{X}, \mathscr{B})$ dominated by $\mathrm{Cap}_{\mathbf{M}}$. Then it is known from [120] that

$$\mathrm{Cap}_{\mathbf{M}}(A) = \overline{\mu}_{\mathbf{M}}(A) = \sup_{\mu \in \mathscr{P}_{\mathbf{M}}} \mu(A). \tag{8.8}$$

8.3.2 The Generalised Bayes Rule for Stochastic Reachability

The construction of the 'global' probability measure $\mathbb{P}$ defined on the sample probability space of $\mathbf{M}$ allows us to replace the reach set probabilities by

$$\mathbb{P}(T_A < T) \quad \text{or} \quad \mathbb{P}(T_A < \infty),$$

where $A \in \mathscr{B}$ is a target set and $T > 0$. In this way, the reachability problem is related with the computation of the capacities associated to the processes $\mathbf{M}_T$ and $\mathbf{M}$, where $\mathbf{M}_T$ is the process $\mathbf{M}$ 'killed' after the time T, i.e.,

$$\mathbb{P}(Reach_T(A)) = \mathrm{Cap}_{\mathbf{M}_T}(A),$$
$$\mathbb{P}(Reach_\infty(A)) = \mathrm{Cap}_{\mathbf{M}}(A).$$

Then the computation of the reach set probabilities associated to A reduces to calculating the capacity of A. Since $\mathrm{Cap}_{\mathbf{M}}$ is a submodular Choquet capacity, Theorem 8.1 (Bayes' theorem for capacities) is applicable under some mild conditions as follows.

Let us suppose that:

- Each Wiener probability $\mathbb{P}_x$ has a density $p(x|\omega)$ with respect to some σ-finite measure on $(\Omega, \mathscr{F})$. Then we define the likelihood function $L(x) = p(x|\omega)$ (constructed as in Sect. 8.2.2). Considering the time horizon $T > 0$, we will also consider the likelihood function $L_T(x) = p(\omega_T, x)$ for x having observed ω_T (the evolution process until T).
- $\mathscr{P} = \mathscr{P}_{\mathbf{M}}$ (respectively $\mathscr{P}_T = \mathscr{P}_{\mathbf{M}_T}$) is an m-closed convex family of probability measures (i.e., a credal set) on $(\mathbf{X}, \mathscr{B})$ dominated by $\mathrm{Cap}_{\mathbf{M}}$ (respectively $\mathrm{Cap}_{\mathbf{M}_T}$).

Following [81], for each $\mu \in \mathscr{P}$ there exists a measurable function f_μ, with $0 \leq f_\mu \leq 1$ such that

$$\mu(f_\mu) = \mathrm{Cap}_{\mathbf{M}}(f_\mu) \tag{8.9}$$

and for each bounded measurable function f there exists $\mu_f \in \mathscr{P}$ such that

$$\mu_f(f) = \mathrm{Cap}_{\mathbf{M}}(f). \tag{8.10}$$

Similar results are true for any $\mu \in \mathscr{P}_T$.

Applying Theorem 8.1 to $\mathrm{Cap}_{\mathbf{M}}$ (respectively $\mathrm{Cap}_{\mathbf{M}_T}$), we get the following estimations:

$$\mathrm{Cap}_{\mathbf{M}}(A|\omega) = \frac{\overline{\mathbb{E}}(L_A)}{\overline{\mathbb{E}}(L_A) + \underline{\mathbb{E}}(L_{A^c})}, \tag{8.11}$$

$$\mathrm{Cap}_{\mathbf{M}}(A|\omega_T) = \frac{\overline{\mathbb{E}}_T(L_{T,A})}{\overline{\mathbb{E}}_T(L_{T,A}) + \underline{\mathbb{E}}_T(L_{T,A^c})}, \tag{8.12}$$

where $\overline{\mathbb{E}}_T$ is the upper expectation defined with respect to $\mathscr{P}_T$. Note that

$$\mathrm{Cap}_{\mathbf{M}}(L_A) = \int p(x|\omega) I_A(x) \mu_{L_A}(dx),$$

where μ_{L_A} is defined as in (8.10).

The formulae (8.11) and (8.12) prompt an idea to introduce expressions for *conditional reach set probabilities* (i.e., the probabilities to reach A having observed the trajectory ω_T or ω)

$$\mathbb{P}\big(Reach_T(A|\omega_T)\big) = \mathrm{Cap}_{\mathbf{M}_T}(A|\omega_T),$$

$$\mathbb{P}\big(Reach_\infty(A|\omega)\big) = \mathrm{Cap}_{\mathbf{M}}(A|\omega). \tag{8.13}$$

In the following we consider only the reachability problem in infinite time horizon and, respectively, only the capacity $\mathrm{Cap}_{\mathbf{M}}$. The case of the reachability problem with finite time horizon can be treated in a similar way taking the process $\mathbf{M}_T$. The 'conditional capacity' formula (8.11) can be extended for the case where we have a set of trajectories E observed (after a 'learning process'). Then we get

$$\begin{aligned}
\mathbb{P}\big(Reach_\infty(A|E)\big) &= \mathrm{Cap}_{\mathbf{M}}(A|E) \\
&= \frac{\overline{\mathbb{E}}(L_{A,E})}{\overline{\mathbb{E}}(L_{A,E}) + \underline{\mathbb{E}}(L_{A^c,E})} \\
&= \frac{\mathrm{Cap}_{\mathbf{M}}^u(L_{A,E})}{\mathrm{Cap}_{\mathbf{M}}^u(L_{A,E}) + \mathrm{Cap}_{\mathbf{M}}^l(L_{A^c,E})},
\end{aligned} \tag{8.14}$$

where

$$L_{A,E}(x) := p(x|\omega) I_A(x) I_E(\omega)$$

and

$$L_{A^c,E}(x) := p(x|\omega) I_{A^c}(x) I_E(\omega).$$

8.3.3 Computing Reach Set Probabilities

Let us consider an agent as a stochastic hybrid system, H with the realisation described by a Markov process $\mathbf{M}$. Suppose that the standard assumptions are satisfied. In the rest of this section we propose some well-established statistical algorithms that might be employed to compute the reach set probabilities.

Solution 1   Let consider that we have given a credal set $\mathcal{K}$, which contains all probability densities of random variables $x_t, t \geq 0$. For a target set $A \in \mathcal{B}$, we define the measurable function

$$\varphi := \sup_{t \geq 0} I_A(x_t). \tag{8.15}$$

Consider μ, an arbitrary element of $\mathcal{K}$, and $\underline{\mu}, \overline{\mu}$ given by (8.4) and (8.3), respectively. If $\underline{\mu}(E) > 0$, where E is an event in the sample probability space (a set of trajectories), the value of $\overline{\mu}(\varphi|E)$ can be computed by the generalised Bayes rule (first proposed in [227]):

$\overline{\mu}(\varphi|E)$ is the unique value of ρ such that $\overline{\mu}\big[(\varphi - \rho)I_E\big] = 0$.

In this case, one can apply Lavine's algorithm, which is a bracketing scheme applied to the generalised Bayes rule, whose objective is to compute upper expectations [162]. Define

$$\underline{\rho}_0 = \inf \varphi I_E$$

and

$$\overline{\rho}_0 = \sup \varphi I_E.$$

Then define

$$m(\rho) = \overline{\mu}\big[(\varphi - \rho)I_E\big].$$

Note that $m(\rho)$ must attain zero in the interval $[\underline{\rho}_0, \overline{\rho}_0]$. Now bracket this interval by repeating the algorithm (for $i \geq 0$):

1. stop if $|\overline{\rho}_i - \underline{\rho}_i| < \varepsilon$ for some positive value ε; or
2. choose ρ_i in $(\underline{\rho}_i, \overline{\rho}_i)$ and, if $m(\rho_i) > 0$, take $\underline{\rho}_{i+1} = \rho_i$ and $\overline{\rho}_{i+1} = \overline{\rho}_i$; if $m(\rho_i) < 0$, take $\underline{\rho}_{i+1} = \underline{\rho}_i$ and $\overline{\rho}_{i+1} = \rho_i$.

The value $m(\rho_i)$ can also provide information on when to stop the bracketing iteration [61].

   The main inconvenience in applying Lavine's algorithm is that function (8.15) is hard to compute. Another iteration scheme, also based on the generalised Bayes' rule, has been proposed by Walley [227] (Note 6.4.1). Both Lavine's and Walley's algorithms have linear convergence.

Solution 2   We consider that the agent H has some beliefs with respect to reach set probabilities, described by an m-closed convex family $\mathcal{P}$ of probability measures on $(\mathbf{X}, \mathcal{B})$ dominated by $\mathrm{Cap_M}$. It should be clear that the credal set $\mathcal{P}$ depends on the random variables $x_t, t \geq 0$, which compose process $\mathbf{M}$. For example, $\mathcal{P}$ could contain the transition probabilities defined by (8.6).

   Since the Choquet capacity associated to $\mathbf{M}$ is the upper envelope of the family of probability measures dominated by $\mathrm{Cap_M}$, the conditional reach set probability (8.13) is, in fact, obtained by conditioning the upper probability $\overline{\mu}_\mathbf{M}$ defined by (8.8).

Credal sets represented by supermodular (and dual by submodular) capacities have been studied in the literature (see [52] and references therein). These credal sets have closed-form expressions for upper posterior probabilities. The algorithms, developed in [52], to compute the conditional upper (respectively lower) posterior probabilities are based on the Möbius transform of an upper (respectively lower) probabilities. These algorithms seem to be the most suitable candidates which can be used to compute the reach set probabilities.

There is a considerable debate about how an agent makes a decision using sets [21, 121]. We consider that a stochastic hybrid system agent should optimise the expected utility, an approach known in robust statistics under the name Γ-minimax [121].

8.4 Some Remarks

In this chapter, we connected stochastic reachability analysis with the Bayesian framework of imprecise probabilities.

1. We provided a Bayesian framework for stochastic hybrid automata, which allows us to characterise the corresponding stochastic reachability problem in terms of upper probabilities or Choquet capacities.
2. Then for the problem of computation of reach set probabilities we found new solutions using different Bayesian inference algorithms already studied in the literature.

Note that this chapter is a 'research chapter', in the sense that in it we initiated the setting for a Bayesian approach to stochastic reachability analysis. Ways to continue this research abound, and they fairly depend on how reach set probabilities are connected to imprecise probabilities.

Chapter 9
Stochastic Reachability Based on Probabilistic Bisimulation

9.1 Overview

In previous chapters, we discussed how to deal with stochastic reachability using probabilistic and analytic methods. In this chapter, we look at this problem from a completely different perspective. The main idea is expressed as follows: Given a process that is too difficult to analyse, we investigate whether there exist other processes that behave in a similar way, but which are much easier to analyse. Therefore, we are looking for a sort of equivalence between systems that preserve some properties of interest, but we abstract away those which are not of interest. This sort of equivalence is usually called *bisimulation*. The concept of bisimulation has been developed mostly in different branches of computer science like concurrency theory, artificial intelligence and category theory.

In computer science, a bisimulation between transition systems is defined as a binary relation between their state spaces, associating systems that behave in the same way in the sense that one system simulates the other. Intuitively, two systems are bisimilar if they mimic each other's moves. If this is the case, an observer cannot distinguish between the evolutions of two bisimilar systems.

In concurrency theory, bisimulation equality, called *bisimilarity*, is the most studied form of behavioural equality for processes and is used in a wide range of applications.

Bisimulation has been discovered not only in computer science but also in other fields like philosophical logic and set theory. Mainly, the bisimulation concept has been derived through refinements of notions of *morphism* between algebraic structures. It is well known that morphisms are maps that are 'structure-preserving'. This concept is essential in all mathematical theories in which the objects of study have some kind of algebraic structure. A morphism provides a way to embed a structure (source) into another structure (target). An *epimorphism* (morphism that is onto) shrinks all the elements and relations existing in the source to analogous items in the target. A *monomorphism* (morphism that is one-to-one) illustrates how the source structure can be viewed as a substructure of the target. If a morphism is both epi and mono, i.e., is an isomorphism, then the two structures are algebraically identical.

L.M. Bujorianu, *Stochastic Reachability Analysis of Hybrid Systems*,
Communications and Control Engineering,
DOI 10.1007/978-1-4471-2795-6_9, © Springer-Verlag London Limited 2012

The search for what lies between morphism and isomorphism led to the discovery of bisimulation.

Probabilistic bisimulation for stochastic systems is usually defined as an equivalence relation between their state spaces that preserves a desired stochastic structure (transition function, expectations, hitting probabilities, excessive functions and so on). We start with the existing classical definitions of stochastic equivalence between Markov processes, then present the meaning of probabilistic bisimulation in certain communities (computer science and control engineering). The quest for notions of equivalence refining the concept of stochastic equivalence led to various new ways to introduce and explore probabilistic/stochastic bisimulation. Moreover, it is possible to further leverage different methodologies to define stochastic bisimulation aiming to obtain stochastic abstractions. Such abstractions are not equivalent to the initial processes: They preserve only some properties we need to perform an analysis of a particular process.

Objectives   In this chapter, our objectives are:

1. to present different concepts of stochastic equivalence;
2. to give a brief exposure of the classical notions of bisimulation for Markov processes;
3. to refine the concept of bisimulation for Markov processes so bisimulation is characterised by a pseudo-metric between processes;
4. to connect stochastic reachability with stochastic bisimulation.

9.2 Stochastic Equivalence

In the following we consider what it means for two stochastic processes to be 'equal'. In the presence of a probability measure, there are three distinct but related ways to define equality between stochastic processes.

Suppose that stochastic processes $\mathbf{M} = (x_t, \mathbb{P}_x)$ and $\widetilde{\mathbf{M}} = (\widetilde{x}_t, \widetilde{\mathbb{P}}_x)$ are defined on the same probability space $(\Omega, \mathscr{F}, \mathbb{P})$. We say that $\widetilde{\mathbf{M}}$ is a *version*, or *modification*, *of* $\mathbf{M}$ if

$$\mathbb{P}(x_t = \widetilde{x}_t) = \mathbb{P}\{\omega | x_t(\omega) = \widetilde{x}_t(\omega)\} = 1$$

for all times $t \geq 0$.

We say that $\mathbf{M}$ and $\widetilde{\mathbf{M}}$ are *indistinguishable* if

$$\mathbb{P}(x_t = \widetilde{x}_t, \forall t \geq 0) = \mathbb{P}\{\omega | x_t(\omega) = \widetilde{x}_t(\omega), \ t \geq 0\} = 1.$$

If two processes are indistinguishable, then they are trivially versions of each other. However, the distinction between 'version' and 'indistinguishable' can be subtle, since two processes can be versions of each other, but exhibiting completely different sample paths.

In order to define both version and indistinguishable, we needed to define both $\mathbf{M}$ and $\widetilde{\mathbf{M}}$ on the same probability space. For our final definition of equivalent stochastic processes, this is not necessary.

Two Markov processes that share the same state space $\mathbf{X}$ and have the same transition probabilities are called *equivalent*. If the processes $\mathbf{M} = (x_t, \mathbb{P}_x)$ and $\widetilde{\mathbf{M}} = (\widetilde{x}_t, \widetilde{\mathbb{P}}_x)$ are equivalent then they have the *same finite-dimensional distributions*, i.e., for any sequence of times $0 < t_1 < t_2 < \cdots < t_n$ and any $\Gamma_1, \Gamma_2, \ldots, \Gamma_n \in \mathscr{B}(\mathbf{X})$

$$\widetilde{\mathbb{P}}_x(\widetilde{x}_{t_1} \in \Gamma_1, \ldots, \widetilde{x}_{t_n} \in \Gamma_n) = \mathbb{P}_x(x_{t_1} \in \Gamma_1, \ldots, x_{t_n} \in \Gamma_n).$$

The question is: Given a Markov process $\mathbf{M}$, what are the transformations of such a process into another equivalent process? As well, one might examine how 'different' are two equivalent processes.

Let $\mathbf{M} = (\Omega, \mathscr{F}_t, x_t, \mathbb{P}_x)$ be an arbitrary Markov process with the sample space Ω. If $\widetilde{\Omega}$ is an arbitrary set and $F : \widetilde{\Omega} \to \Omega$ satisfies the condition

$$A \supseteq F(\widetilde{\Omega}) \Rightarrow \mathbb{P}_x(A) = 1,$$

we may set up a new process

$$\widetilde{x}_t(\widetilde{\omega}) := x_t\big(F(\widetilde{\omega})\big), \quad t \geq 0,$$
$$\widetilde{\mathscr{F}}_t = F^{-1}(\mathscr{F}_t), \quad t \geq 0,$$
$$\widetilde{\mathbb{P}}_x\big(F^{-1}(A)\big) = \mathbb{P}_x(A), \quad A \in \mathscr{F}_0.$$

It can be proved (see [77]) that the new process $\widetilde{\mathbf{M}} = (\widetilde{\Omega}, \widetilde{\mathscr{F}}_t, \widetilde{x}_t, \widetilde{\mathbb{P}}_x)$. We say that the process $\widetilde{\mathbf{M}}$ is obtained from $\mathbf{M}$ by means of the transformation F of the sample space.

According to the same reference [77], two particular cases of this methodology are to be recorded:

- *Weeding out the sample space* Ω: The transformation F is chosen to be one-to-one (injective), and then the space $\widetilde{\Omega}$ is identified with a subspace of Ω. In this case, the random variables x_t are restricted to $\widetilde{\Omega}$, the elements of $\mathscr{F}_t$ are replaced by their intersections with $\widetilde{\Omega}$ and, accordingly, the Wiener probabilities $\mathbb{P}_x$ are restricted to these intersections.
- *Splitting the sample space* Ω: The transformation F is chosen to be onto (surjective), and then each point $\omega \in \Omega$ is viewed as being split into the set of points $F^{-1}(\omega) = \{\widetilde{\omega} \mid F(\widetilde{\omega}) = \omega\}$.

Simple reasoning leads to the fact that any transformation of the sample state space can be reduced to successive operations of weeding and splitting of the elementary events $\omega \in \Omega$.

Given a Markov process, other stochastic transformations include:

- time changes using additive functionals,
- measure changes of the Wiener probabilities,
- killing/stopping using multiplicative functionals.

In principle, such transformations preserve somehow the trajectories of the initial process. In the next section, we introduce the idea of bisimulation or 'lumpability', which aims to lump together those states (and trajectories) that show similar behaviour. The final goal of bisimulation is to simplify the state space and the dynamics of a stochastic process so it becomes much easier to analyse.

9.3 Stochastic Bisimulation

Suppose we have a (discrete-time or continuous-time) transient Markov process (x_t) with the state space $\mathbf{X}$ and transition probabilities (p_t). Suppose further that $\mathbf{X}$ is a Polish or analytic space and is equipped with its Borel σ-algebra $\mathscr{B}(\mathbf{X})$ or $\mathscr{B}$.

A subset A of $\mathbf{X}$ is said to be *analytic* if there exist a Borel space Z and a Borel subset B of $\mathbf{X} \times Z$ such that $A = proj_{\mathbf{X}}(B)$, where $proj_{\mathbf{X}}$ is the projection mapping from $\mathbf{X} \times Z$ to $\mathbf{X}$. If A is analytic, $\mathbf{X} \backslash A$ is not necessarily analytic (see [22]). It is clear that every Borel subset of a Borel space $\mathbf{X}$ is also an analytic subset of $\mathbf{X}$.

There are at least three natural σ-algebras in a Borel space $\mathbf{X}$. The first is the Borel σ-algebra $\mathscr{B}$ mentioned earlier. The second is the σ-algebra generated by the analytic subsets of $\mathbf{X}$, called the *analytic σ-algebra* and denoted by $\mathscr{A}$. The third is the *universal σ-algebra* $\mathscr{U}$, which is the intersection of all completions of $\mathscr{B}$ with respect to all probability measures. Thus, $\Gamma \in \mathscr{U}$ if and only if, given any probability measure p on $(\mathbf{X}, \mathscr{B})$, there is a Borel set B and a p-null set $\mathscr{N}$ such that $\Gamma = B \cup \mathscr{N}$. It is known [22] that

$$\mathscr{B} \subset \mathscr{A} \subset \mathscr{U}.$$

Any probability measure p on $(\mathbf{X}, \mathscr{B})$ has a unique extension to a probability measure $\overline{p}$ on $(\mathbf{X}, \mathscr{U})$. We write p instead of $\overline{p}$.

In the previous section, we pointed out those equivalences defined with respect to the trajectories of the stochastic processes. Reasoning about trajectories is not always simple. Often, in practice, given a complex stochastic process, we aim to find another process that mimics the behaviour of the first, but is simpler to analyse. One way to simplify a stochastic process is to 'reduce' its state space. One way to approach reducing the state space complexity is to define a state space transformation onto a simpler state space. The reducing method was introduced in the early 1960s, and it received rigorous exposition in [77]. Since then it has been the classical approach to state space simplification. The simplest way to define a state space transformation is via an equivalence relation and its associated projection map. This relation should 'preserve' the transition probabilities. The quotient process will be equivalent to the given one, in the sense that its trajectories will represent equivalent classes with respect to the induced equivalence relation on the trajectories of the initial process. Usually, such an equivalence relation is referred to as *probabilistic* or *stochastic bisimulation*.

If $\mathscr{R}$ is a binary relation on $\mathbf{X}$, a subset $A \subseteq \mathbf{X}$ is closed under $\mathscr{R}$ if

$$\{y \in \mathbf{X} | \exists x \in A \text{ such that } x \mathscr{R} y\} \subseteq A.$$

The σ-algebra of universally measurable $\mathscr{R}$-closed sets is denoted by $\mathrm{cl}_{\mathscr{U}}(\mathscr{R})$.

A (strong) bisimulation relation is defined as an equivalence relation $\mathscr{R} \subset \mathbf{X} \times \mathbf{X}$ such that

$$x \mathscr{R} y \quad \text{iff} \quad p_t(x, A) = p_t(y, A) \tag{9.1}$$

for all $\mathscr{R}$-closed measurable sets A of $\mathbf{X}$. Let $\mathbf{X}/_{\mathscr{R}}$ be the quotient space, i.e., the space of equivalence classes of $\mathbf{X}$ with respect to $\mathscr{R}$. We denote the equivalence class of $x \in \mathbf{X}$ with respect to $\mathscr{R}$ by $[x]$.

Let $\pi : \mathbf{X} \to \mathbf{X}/\mathscr{R}$, $\pi x := [x]$ and define on $\mathbf{X}/\mathscr{R}$ the topology induced by π. Let $\mathscr{B}(\mathbf{X}/\mathscr{R})$ be the Borel σ-algebra of $\mathbf{X}/\mathscr{R}$. In fact, $\mathscr{B}(\mathbf{X}/\mathscr{R})$ is composed by $\mathscr{R}$-closed measurable sets of $\mathbf{X}$. Then the relation (9.1) ensures that we can define the transition probabilities of the quotient process as

$$\widehat{p}_t\big([x], A\big) = p_t(x, A), \quad x \in \mathbf{X}, \ A \in \mathscr{B}(\mathbf{X}/\mathscr{R}), \ t \geq 0 \tag{9.2}$$

or

$$\widehat{p}_t(\pi x, A) = p_t\big(x, \pi^{-1}(A)\big), \quad x \in \mathbf{X}, \ A \in \mathscr{B}(\mathbf{X}/\mathscr{R}), \ t \geq 0. \tag{9.3}$$

When a bisimulation relation is available for a stochastic process, the equality of the finite-dimensional distributions of the given process and its quotient process holds only with respect to $\mathscr{R}$-closed measurable sets of $\mathbf{X}$.

The transition probabilities are stochastic kernels (for each $t \geq 0$, $p_t : \mathbf{X} \times \mathscr{B} \to [0, 1]$), which serve to introduce some operators on the space of bounded measurable real-valued functions on $\mathbf{X}$, denoted by $\mathscr{B}^b(\mathbf{X})$. The operator semigroup (P_t) maps $\mathscr{B}^b(\mathbf{X})$ to itself and is given by

$$(P_t f)(x) = \int f(y) p_t(x, dy) = \mathbb{E}_x\big[f(x_t)\big].$$

The kernel operator $V : \mathscr{B}^b(\mathbf{X}) \to \mathscr{B}^b(\mathbf{X})$ is given by

$$Vf(x) = \int_0^\infty (P_t f)(x) \, dt.$$

A class monotone argument allows us to write (9.3) as

$$\widehat{P}_t \widehat{f} \circ \pi = P_t(\widehat{f} \circ \pi), \quad \widehat{f} \in \mathscr{B}^b(\mathbf{X}/\mathscr{R}), \quad t \geq 0, \tag{9.4}$$

or

$$\widehat{V} \widehat{f} \circ \pi = V(\widehat{f} \circ \pi), \quad \widehat{f} \in \mathscr{B}^b(\mathbf{X}/\mathscr{R}), \tag{9.5}$$

where $(\widehat{P}_t)$ and $\widehat{V}$ correspond to the quotient process.

Let us define a function $\pi^* : \mathscr{B}^b(\mathbf{X}/\mathscr{R}) \to \mathscr{B}^b(\mathbf{X})$ given by $\pi^* \widehat{f} = \widehat{f} \circ \pi$. Then formula (9.4) becomes

$$\pi^* \circ \widehat{P}_t = P_t \circ \pi^*. \tag{9.6}$$

The relation (9.6) is known as the *Dynkin intertwining relation*. This implies that the finite-dimensional distributions of $\pi \circ \mathbf{M}$ under $\mathbb{P}_x$ are the same as those of $\widehat{\mathbf{M}}$ under $\widehat{P}^{\pi x}$ for any $x \in \mathbf{X}$ (where $\mathbf{M}$ is the initial process and $\widehat{\mathbf{M}}$ is the quotient process). The relation (9.6) says that π is a *Markov function*.

Using the connection between the kernel operator and the infinitesimal generator of a Markov process, the relation (9.5) is equivalent to

$$\pi^* \circ \widehat{L} = L \circ \pi^*,$$

where L and $\widehat{L}$ are, respectively, the generator of $\mathbf{M}$ and $\widehat{\mathbf{M}}$.

9.4 Bisimulation via a Stochastic Kernel

Consider $\mathscr{R}$ an equivalence relation on $\mathbf{X}$ and $\pi : \mathbf{X} \to \mathbf{X}/_{\mathscr{R}}$ the corresponding projection. Let $U : \mathbf{X}/_{\mathscr{R}} \times \mathscr{B}(\mathbf{X}) \to [0, 1]$ be a stochastic kernel such that

$$U\big([x], \pi^{-1}([x])\big) = 1. \tag{9.7}$$

This is the *distributor kernel*. We define the probability transitions of the quotient process as

$$\widehat{p}_t\big([x], A\big) = \int p_t\big(y, \pi^{-1}(A)\big)U\big([x], dy\big). \tag{9.8}$$

Then we can define the operator $\mathbf{U} : \mathscr{B}^b(\mathbf{X}) \to \mathscr{B}^b(\mathbf{X}/_{\mathscr{R}})$ by

$$(\mathbf{U}f)\big([x]\big) = \int_{\mathbf{X}} f(y)U\big([x], dy\big).$$

A simple calculation gives

$$\mathbf{U}(\widehat{f} \circ \pi)\big([x]\big) = \int (\widehat{f} \circ \pi)(y)U\big([x], dy\big)$$

$$= \int_{\pi^{-1}([x])} (\widehat{f} \circ \pi)(y)U\big([x], dy\big)$$

$$= \widehat{f}([x])$$

i.e., $\mathbf{U}(\widehat{f} \circ \pi) = \widehat{f}$ or

$$\mathbf{U}\pi^* = I, \tag{9.9}$$

where I is the identity of $\mathscr{B}^b(\mathbf{X}/_{\mathscr{R}})$.

Remark 9.1 If there exists a density function u of U (the Radon–Nikodym derivative of U with respect to a probability measure m on $\mathbf{X}$), i.e.,

$$U\big([x], E\big) = \int u\big([x], y\big)m(dy),$$

then

$$\widehat{p}_t\big([x], A\big) = \int p_t\big(y, \pi^{-1}(A)\big)u\big([x], y\big)m(dy).$$

Proposition 9.2 $\mathscr{R}$ *is a strong bisimulation if and only if*

$$\pi^* \mathbf{U} P_t \pi^* = P_t \pi^*. \tag{9.10}$$

Using (9.8), one can compute the quotient process semigroup as

$$\widehat{P}_t \widehat{f} = \mathbf{U}\big(P_t \pi^* \widehat{f}\big).$$

Remark 9.3 Condition (9.10) is equivalent to (the Dynkin intertwining condition)

$$\pi^* \widehat{P}_t = P_t \pi^*.$$

Proposition 9.4 *A sufficient condition for $\mathscr{R}$ to be a bisimulation of $\mathbf{M}$ with the initial probability distribution equal to μ is that*

$$\mathbf{U} P_t \pi^* \mathbf{U} = \mathbf{U} P_t, \tag{9.11}$$

where $\mathbf{U}$ satisfies a 'compatibility' relation with respect to μ,

$$\int \pi^* (\mathbf{U} f)(x) \mu(dx) = \int f(x) \mu(dx), \quad \forall f \in \mathscr{B}^b(\mathbf{X}). \tag{9.12}$$

Remark 9.5 Condition (9.12) is equivalent to

$$\mu(E) = \int U\big(\pi x, E \cap \pi^{-1}(\pi x)\big) \mu(dx)$$

for any measurable Borel set $E \in \mathscr{B}(\mathbf{X})$.

The initial probability distribution $\widehat{\mu}$ of the quotient process has to satisfy

$$\int (\mathbf{U} f)([x]) \widehat{\mu}(d[x]) = \int f(x) \mu(dx); \tag{9.13}$$

i.e., for any measurable Borel set $E \in \mathscr{B}(\mathbf{X})$

$$\mu(E) = \int U\big([x], E \cap \pi^{-1}([x])\big) \widehat{\mu}(d[x]).$$

Remark 9.6 Condition (9.11) is equivalent to

$$\mathbf{U} P_t = \widehat{P}_t \mathbf{U} \tag{9.14}$$

and conditions (9.7), (9.9) and (9.14) represent the Pitman and Rogers conditions for π to be a Markov function.

Remark 9.7 Clearly, for an element $[x] \in \mathbf{X}/\mathscr{R}$

$$\mu = U\big([x], \cdot\big), \qquad \widehat{\mu} = \delta_{[x]}$$

satisfy the conditions (9.12) and (9.13).

9.5 Bisimulation Between Two Stochastic Processes

In this section, we propose ways to relax of the concept of strong stochastic bisimulation presented in the previous section. In practice, it is hard to check the equality of transition probabilities for two stochastic processes. Moreover, two processes may exhibit similar behaviour even if this equality is not fulfilled. It might be the case that two processes behave in a similar way if either

- other statistical parameters (expectations, moments, variations, capacities) are approximately equal, or
- it is possible to define certain metrics between processes that become zero in case of bisimulation.

From a mathematical perspective, the first research direction means finding ways to reduce the order of the forward or backward Kolmogorov equations of the given process. This sort of approach might be difficult and lengthy and it may be not suitable for practical implementation. The second approach seems to be more appealing for researchers in the field. Given two stochastic processes, there exist suitable ways to define metrics or bisimulation functions that can be used in various criteria for checking bisimilarities of the underlying processes. This research direction is still under development and the most notable references are [138–140].

In the following, we sketch one way one to define weak bisimulation that might be useful for stochastic reachability analysis. This method is based on capacities associated to Markov processes. Similar methods make use of the infinitesimal generator, Dirichlet forms, (super)martingales and spectral operator theories associated to Markov processes. A theory's efficiency depends on the ability of these theories to characterise the temporal evolutions of the related processes.

Let $(\mathbf{X}, \mathscr{B}(\mathbf{X}))$ and $(\mathbf{Y}, \mathscr{B}(\mathbf{Y}))$ be Polish/analytic spaces and let $\mathscr{R} \subset \mathbf{X} \times \mathbf{Y}$ be a relation such that $\Pi^1(\mathscr{R}) = \mathbf{X}$ and $\Pi^2(\mathscr{R}) = \mathbf{Y}$.

We define the equivalence relation on $\mathbf{X}$ that is induced by the relation

$$\mathscr{R} \subset \mathbf{X} \times \mathbf{Y}$$

as the transitive closure of

$$\left\{ (x, x') \middle| \exists y \text{ s.t.} (x, y) \in \mathscr{R} \text{ and } (x', y) \in \mathscr{R} \right\}.$$

We write $\mathbf{X}/_{\mathscr{R}}$ and $\mathbf{Y}/_{\mathscr{R}}$ for the sets of equivalence classes of $\mathbf{X}$ and $\mathbf{Y}$ induced by $\mathscr{R}$. We denote the equivalence class of $x \in \mathbf{X}$ by $[x]$. We define now the notion of *measurable relation*.

Let

$$\mathscr{B}^*(\mathbf{X}) = \mathscr{B}(\mathbf{X}) \cap \left\{ A \subset \mathbf{X} \middle| x \in A \text{ and } [x] = [x'] \Rightarrow x' \in A \right\}$$

be the collection of all Borel sets, in which any equivalence class of $\mathbf{X}$ is either totally contained in or totally outside. It can be checked that $\mathscr{B}^*(\mathbf{X})$ is a σ-algebra. Let $\mathbf{X}/_{\mathscr{R}}$ be the set of equivalence classes of $\mathbf{X}$; let $\pi_{\mathbf{X}} : \mathbf{X} \to \mathbf{X}/_{\mathscr{R}}$ be the mapping that maps each $x \in \mathbf{X}$ into its equivalence class and let

$$\mathscr{B}(\mathbf{X}/_{\mathscr{R}}) = \left\{ A \subset \mathbf{X}/_{\mathscr{R}} \middle| \pi_{\mathbf{X}}^{-1}(A) \in \mathscr{B}^*(\mathbf{X}) \right\}.$$

Then $(\mathbf{X}/_{\mathscr{R}}, \mathscr{B}(\mathbf{X}/_{\mathscr{R}}))$, which is a measurable space, is called the *quotient space of* $\mathbf{X}$ *with respect to* $\mathscr{R}$. The quotient space of $\mathbf{Y}$ with respect to $\mathscr{R}$ is defined in a similar way.

We define a bijective mapping $\psi : \mathbf{X}/_{\mathscr{R}} \to \mathbf{Y}/_{\mathscr{R}}$ as $\psi([x]) = [y]$ if $(x, y) \in \mathscr{R}$ for some $x \in [x]$ and some $y \in [y]$. We say that the relation $\mathscr{R}$ is *measurable* in $\mathbf{X}$ and $\mathbf{Y}$ if for all $A \in \mathscr{B}(\mathbf{X}/_{\mathscr{R}})$ we have $\psi(A) \in \mathscr{B}(\mathbf{Y}/_{\mathscr{R}})$ and vice versa, i.e., ψ

is a homeomorphism. Then the real measurable functions defined on $\mathbf{X}/_{\mathcal{R}}$ can be identified with those defined on $\mathbf{Y}/_{\mathcal{R}}$ through the homeomorphism ψ. We can write

$$\mathcal{B}^b(\mathbf{X}/_{\mathcal{R}}) \stackrel{\psi}{\cong} \mathcal{B}^b(\mathbf{Y}/_{\mathcal{R}}).$$

Moreover, these functions can be thought of as real functions defined on $\mathbf{X}$ or $\mathbf{Y}$ measurable with respect to $\mathcal{B}^*(\mathbf{X})$ or $\mathcal{B}^*(\mathbf{Y})$.

Definition 9.8 Suppose we have the capacities $c_{\mathbf{X}}$ and $c_{\mathbf{Y}}$ defined on the analytic spaces $(\mathbf{X}, \mathcal{B}(\mathbf{X}))$ and $(\mathbf{Y}, \mathcal{B}(\mathbf{Y}))$, respectively. Suppose that we have a measurable relation $\mathcal{R} \subset \mathbf{X} \times \mathbf{Y}$. The capacities $c_{\mathbf{X}}$ and $c_{\mathbf{Y}}$ are called *equivalent with respect to* $\mathcal{R}$ if they define the same capacity on the quotient space of $\mathbf{X}$ and $\mathbf{Y}$, i.e., if

$$c_{\mathbf{X}}\big(\pi_{\mathbf{X}}^{-1}(A)\big) = c_{\mathbf{Y}}\big(\pi_{\mathbf{Y}}^{-1}[\psi(A)]\big)$$

for all $A \in \mathcal{B}(\mathbf{X}/_{\mathcal{R}})$.

Suppose we have two Borel right Markov processes $\mathbf{M}$ and $\mathbf{W}$ with state spaces $\mathbf{X}$ and $\mathbf{Y}$, respectively.

Definition 9.9 A measurable relation $\mathcal{R} \subset \mathbf{X} \times \mathbf{Y}$ is a *bisimulation between* $\mathbf{M}$ *and* $\mathbf{W}$ if their associated capacities $\mathrm{Cap}_{\mathbf{M}}$ and $\mathrm{Cap}_{\mathbf{W}}$ are equivalent with respect to $\mathcal{R}$.

It is known that if two processes are symmetric and are defined on the same state space, the equality of their capacities implies that these processes are time changes of one to another [89].

We can define now a *pseudo-metric with respect a measurable relation* $\mathcal{R} \subset \mathbf{X} \times \mathbf{Y}$ between the processes $\mathbf{M}$ and $\mathbf{W}$ as follows:

$$d_{\mathcal{R}}(\mathbf{M}, \mathbf{W}) = \sup_{f \in \mathcal{B}^{*b}(\mathbf{X})} \left| \int f \, d\mathrm{Cap}_{\mathbf{M}} - \int f \circ \psi \, d\mathrm{Cap}_{\mathbf{W}} \right|$$

where $\mathcal{B}^{*b}(\mathbf{X})$ is the set of bounded real $\mathcal{B}^*(\mathbf{X})$-measurable functions on $\mathbf{X}$.

Remark 9.10 We can define a distance between two processes if and only there exists a relation on the product of their state spaces $\mathbf{X} \times \mathbf{Y}$ such that the two quotient spaces are homeomorphic. Or, equivalently, if there exist a third measurable space $(Z, \mathcal{B}(Z))$ and two surjective measurable mappings $\phi_1 : \mathbf{X} \to Z$ and $\phi_2 : \mathbf{Y} \to Z$, then

$$d(\mathbf{M}, \mathbf{W}) = \sup_{f \in \mathcal{B}^b(Z)} \left| \int f \circ \phi_1 \, d\mathrm{Cap}_{\mathbf{M}} - \int f \circ \phi_2 \, d\mathrm{Cap}_{\mathbf{W}} \right|,$$

where $\mathcal{B}^b(Z)$ is the set of bounded real $\mathcal{B}^b(Z)$-measurable functions on Z.

Proposition 9.11 A measurable relation $\mathcal{R} \subset \mathbf{X} \times \mathbf{Y}$ is a bisimulation between $\mathbf{M}$ and $\mathbf{W}$ if and only if

$$d_{\mathcal{R}}(\mathbf{M}, \mathbf{W}) = 0.$$

In the classical theory of stochastic processes, two processes are considered equivalent if and only if their transition probabilities differ on a set of times of measure zero.

Proposition 9.12 *Two equivalent processes are bisimilar.*

We can refine further this result by considering much finer ways to measure sets i.e., by considering capacities.

Proposition 9.13 *Two processes that differ on set of times of capacity zero are bisimilar.*

A similar results holds for a set of states.

Proposition 9.14 *Two processes that differ on a set of states of capacity zero are bisimilar.*

We consider now a simplified situation in air traffic control. A stochastic model of commercial flights from London's airports Stansted and Gatwick to Paris is constructed. This model varies periodically, as there are constant changes in weather influence (strong winds, storms, dark clouds, etc.) and local traffic (e.g., domestic flights from Ashford Airport or traffic between Europe and the States). These influences are captured in the definition of various concepts that characterise a Markov process, like generator, resolvent and transition kernels. Often, in such models the set of trajectories starting from Stansted and passing through Kent county (the area marked II on the first map) has the same probability as the set of trajectories starting from Gatwick and passing through Sussex county (the area marked I on the first map). This is because the weather in the two counties is very similar and the local traffic presents the same characteristics.

The Markov process illustrated in Fig. 9.1 is bisimilar to the one illustrated in Fig. 9.2. Bisimilarity is not a refinement game of partitions of a geographical area. The air traffic control system state space is actually given by the Borel sets of the Euclidean space. The reason is that in the ATC representation of a plane position, the precise (latitude and longitude) position is not considered, but rather a larger area is, one that can be measured by a probability. Because of this, it does not really matter if the plane is in fact, for example, 30 meters away from the controller's representation. Here, the two counties are bisimilar from the perspective of weather conditions and the local traffic on some routes.

9.6 Bisimulations Preserving Reach Probabilities

In this section, we investigate which bisimulation/equivalence relations are best suited to facilitate stochastic reachability analysis.

Fig. 9.1 Stochastic bisimilar
air traffic regions

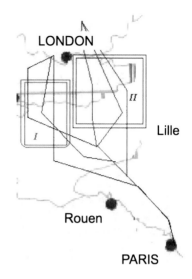

Let us consider

$$\mathbf{M} = (\Omega, \mathscr{F}, \mathscr{F}_t, \theta_t, x_t, \mathbb{P}_x)$$

as the realisation of a stochastic hybrid system H.

For each $x \in \mathbf{X}$, the kernel operator V will provide a measure V_x defined by

$$V_x(A) = V I_A(x), \quad \forall A \in \mathscr{B},$$

and for any measurable positive function f on $\mathbf{X}$ we have

$$Vf(x) = \int f \, dV_x.$$

To address the reachability problem assume that we have a given set $E \in \mathscr{B}(\mathbf{X})$ and a horizon time $T > 0$. Let us consider the reach events $Reach_T(E)$ and $Reach_\infty(E)$.

The reachability problem is related to the computation of the capacities associated to the processes $\mathbf{M}_T$ and $\mathbf{M}$, where $\mathbf{M}_T$ is process $\mathbf{M}$ 'killed' after the time T.

On the other hand, we would like to characterise the sets

$$Reach_T^{\text{init}}(E) = \left\{ x \in \mathbf{X} \,\middle|\, \exists \omega \in \Omega, \exists t \in [0, T] : \phi(t, \omega, x) \in E \right\},$$
$$Reach_\infty^{\text{init}}(E) = \left\{ x \in \mathbf{X} \,\middle|\, \exists \omega \in \Omega, \exists t \in [0, \infty) : \phi(t, \omega, x) \in E \right\},$$

where $\phi(t, \omega, x)$ is a trajectory of $\mathbf{M}$ starting with $x \in \mathbf{X}$. These are thought of as sets of initial points, which give trajectories of $\mathbf{M}$ with nonempty intersection with E.

Lemma 9.15 *For any measurable set $E \in \mathscr{B}$ and for $T > 0$, we have*

$$Reach_T^{\text{init}}(E) = \left\{ x \in \mathbf{X} \,\middle|\, \sup_{t \in [0,T]} \mathbf{P}_t I_E(x) > 0 \right\}.$$

Fig. 9.2 Unification of
bisimilar state regions

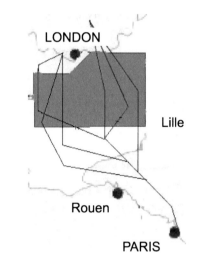

Proposition 9.16 *If* **M** *has the càdlàg property and G is an open set of* **X** *then*

$$Reach_\infty^{init}(G) = \{x \in \mathbf{X}\,|\,V_x(G) > 0\}.$$

Remark 9.17 The measure V_x does not suit our purposes: V_x accounts each trajectory ω every time when it 'visits' E. This weakness is eliminated if we consider the measure $\mathbb{P}(T_E < \infty)$.

Suppose we have are two stochastic hybrid systems H and H' with the realisations **M** and **W** (with the state spaces **X** and **Y**).

Definition 9.18 H and H' are *bisimilar* if there exists a measurable relation $\mathscr{R} \subset$ **X** $\times$ **Y** such that $\mathscr{R}$ is a bisimulation between **M** and **W**.

Proposition 9.19 $\mathscr{R} \subset \mathbf{X} \times \mathbf{Y}$ *is a bisimulation relation between H and H' if and only if the probabilities of reachable events associated to 'saturated' (with respect to $\mathscr{R}$) Borel sets are equal, i.e.,*

$$\mathbb{P}_\mathbf{M}(T_E < \infty) = \mathbb{P}_\mathbf{W}(T_{\psi(E)} < \infty)$$

for all $E \in \mathscr{B}^*(\mathbf{X})$.

The proof is a clear consequence of definition of a bisimulation relation between two Markov processes.

The above proposition shows that our definition of bisimulation between stochastic hybrid systems is natural, since probabilities of reachable events are preserved. Then, naturally, the reachability analysis of a stochastic hybrid system can be performed using much simpler stochastic hybrid systems bisimilar to the given one.

9.7 Some Remarks

In this chapter we have tried to find possible answers to the following question:

How can we define a process that is 'bisimilar' to a given one (more complex) such that the study of the stochastic reachability problem becomes much easier?

Of course, all classical stochastic equivalences between Markov processes preserve the reach set probabilities. The problem is that we do not gain much by replacing a process with an equivalent one. The complexity remains fairly the same. We would like to have 'recipes' to cluster together the states and the trajectories of a given process so that the new process will be much simpler to analyse. This is not an easy problem and there is still ongoing research to find such 'recipes'. It is clear that looking to find processes that have exactly the same desired stochastic parameters is the wrong way to deal with this problem. The natural way to undertake this line of research is to investigate 'approximate bisimulations' between processes [1, 221] or even further approximate abstractions [138]. The interested reader should consult the existing literature regarding such bisimulations/abstractions. A presentation of them here would deviate too far from the main focus of this book.

Chapter 10
Stochastic Reachability with Constraints

10.1 Overview

This chapter is concerned with leveraging the concept of stochastic reachability such that it may capture state and time constraints, dynamics of the target set, randomisation of the time horizon and so on. These represent hot topics in modern stochastic reachability research. Analytic solutions will be provided only in the case of state-constrained stochastic reachability, while existing research avenues will be described for the other extensions of stochastic reachability.

State-constrained stochastic reachability represents the stochastic version of some well-developed problems for deterministic systems, including:

- The 'reach-avoid' problem: determine the set of initial conditions for which one can find at least one control strategy to steer the system to a target set while also satisfying some constraints regarding the avoidance of certain known obstacles.
- The viability problem: determine the set of initial conditions for which one can find at least one control strategy to keep the system in a given set.
- The scheduling problem: determine the set of initial conditions for which one can find at least one control strategy to steer the system to a target set before it visits another state.

In this chapter, we formulate state-constrained stochastic reachability for a stochastic hybrid process with no controllers. The controlled version will be presented in a subsequent volume of this title. Formulation of the problem is nice and intuitive, but finding computational solutions is a challenging task. An analytic method based on PDEs will be presented. To investigate trajectories that go towards a target while satisfying some constraints, the reader more interested in simulations will need characterisations based on other probabilistic concepts, like entropy and probability current. For the discrete case, intuitive representations can be found using the connections between Markov chains and electrical networks.

Objectives The objectives of this chapter can be summarised as follows:

1. To define formally the concept of state-constrained stochastic reachability.

L.M. Bujorianu, *Stochastic Reachability Analysis of Hybrid Systems*,
Communications and Control Engineering,
DOI 10.1007/978-1-4471-2795-6_10, © Springer-Verlag London Limited 2012

2. To provide analytical characterisations of the above concept.
3. To explain the deployment of state-constrained stochastic reachability at the different scales of an SHS model.
4. To give further generalisations of the above concept.

10.2 Mathematical Definitions

In this section, we extend the concept of state-constrained reachability defined in the literature (see [14] and references therein) doing so only for discrete/continuous time Markov chains (with discrete state space) to continuous time/space Markov processes. Further, we study this concept for stochastic hybrid processes.

Let us consider a stochastic hybrid process

$$\mathbf{M} = (x_t, \mathbb{P}_x)$$

with state space $(\mathbf{X}, \mathscr{B}(\mathbf{X}))$. State-constrained reachability analysis denotes a reachability problem with additional conditions (constraints) on system trajectories. Let us consider A, B, two Borel-measurable sets of the state space $\mathbf{X}$ with disjoint closures, i.e.,

$$A, B \in \mathscr{B}(\mathbf{X}) \quad \text{and} \quad \overline{A} \cap \overline{B} = \emptyset.$$

We consider two fundamental situations. Suppose the system paths start from a given initial state x and we are interested in a target state set, say B. These trajectories can hit the state set A or not. Therefore, we may define two new concepts:

- *Obstacle avoidance reachability* (see Fig. 10.1). In this interpretation B is a safe set, while A is not. The goal is to compute the probability, denoted by

$$p^B_{\neg A}(x),$$

of all trajectories that start from a given initial state x and hit the set B without hitting the state set A.
- *Waypoint reachability* (see Fig. 10.2). In this interpretation we are interested in computing the probability, denoted by

$$p^B_A(x),$$

of all trajectories that hit B only after hitting A.

The connection between the two types of stochastic reachability is given by the formula

$$p^B_{\neg A}(x) + p^B_A(x) = \varphi_B(x)$$

where φ_B is the reachability function for the target set B given by formula (7.3). Therefore, computations of the probabilities corresponding to the two types of reachability are equivalent. To have an easy notation, it is more convenient to work

Fig. 10.1 Obstacle
avoidance reachability

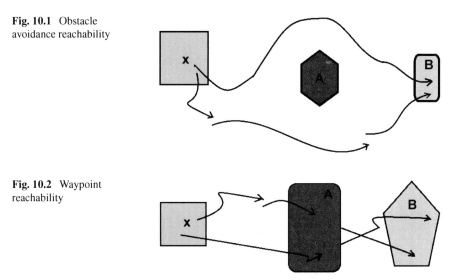

Fig. 10.2 Waypoint
reachability

with the waypoint reachability, which will be called from now on simply *state-constrained reachability*.

Now we consider the executions (paths) of a stochastic hybrid process that starts in $x = (q, z) \in \mathbf{X}$. When we investigate state-constrained reachability, we seek the probability that these trajectories visit A before visiting, eventually, B. Mathematically, this is the probability of

$$\{\omega \,|\, x_t(\omega) \notin B, \forall t \leq T_A\}.$$

Moreover, using the first hitting time T_B of B, we are interested in computing

$$p_A^B(x) = \mathbb{P}_x[T_A < T_B]. \tag{10.1}$$

Consequently, state-constrained reachability is related to some classical topics treated in the literature of Markov processes, like the first passage problem [109], excursion theory [100] and estimation of the equilibrium potential of the capacitor of the two sets [37]. These references provide theoretical characterisations for probabilities (10.1) for different classes of Markov processes. Here, the scope is not to survey all these characterisations but to identify the appropriate analytical solutions of this problem.

10.3 Mathematical Characterisations

The main scope of this section is to prove that state-constrained reachability probabilities can be characterised as solutions of certain boundary value problems expressed in terms of the infinitesimal generator of the given stochastic hybrid process.

We will use the concepts of excessive function and kernel operator V that can be introduced with respect to a Markov process $\mathbf{M}$ (see Chap. 2).

The following assumption is essential for the results of this section.

Assumption 10.1 Suppose that M is a *transient Markov process*, i.e., there exists a strictly positive Borel-measurable function q such that Vq is a bounded function.

Using [37], we have the following characterisation.

Proposition 10.1 *State-constrained reachability probability p_A^B has the following properties*:

(i) $0 \leq p_A^B \leq 1$ *a.s. on* $\mathbf{X}$,
(ii) $p_A^B = 0$ *a.s. on B and $p_A^B = 1$ on A*,
(iii) p_A^B *is the potential of a signed measure v such that the support of v^+ is contained in A and the support of v^- is contained in B.*

We can write, in a more compact manner,

$$p_A^B(x) = \begin{cases} \mathbb{P}_x[T_A < T_B], & \text{if } x \notin A \cup B \\ 1, & \text{if } x \in A \\ 0, & \text{if } x \in B. \end{cases}$$

An inclusion–exclusion argument leads to the following formula:

$$\begin{aligned} p_A^B(x) &= \mathbb{P}_x(T_A < T_B) \\ &= P_A 1(x) - P_B P_A 1(x) + P_A P_B P_A 1(x) - \cdots. \end{aligned}$$

We make the following notation for:

- composition of the hitting operators corresponding to the target sets A and B

$$V^{A \to B} := P_A \circ P_B,$$

where

$$\begin{aligned} P_A(P_B u)(x) &= \mathbb{E}_x\left\{ (P_B u)(x_{T_A}) \right\} \\ &= \mathbb{E}_x\left\{ \mathbb{E}_{x_{T_A}} u(x_{T_B}) \right\}, \quad u \in \mathscr{B}^b(\mathbf{X}), \end{aligned}$$

provided that $T_A, T_B < \zeta$ (ζ is the lifetime of the process);
- the probability of hitting A again after n excursions between A and B

$$p_n := \left(V^{A \to B} \right)^n \varphi_A,$$

where φ_A is given by (7.3);
- the probability of hitting A again after 'infinitely many' excursions between A and B

$$\Gamma := \sum_{n=0}^{\infty} p_n.$$

Proposition 10.2 *Then we have the following recurrence formula*:

$$p_A^B = (I - P_B)\Gamma$$

where $I : \mathscr{B}^b(\mathbf{X}) \to \mathscr{B}^b(\mathbf{X})$ *is the identity operator.*

Proof Each p_n is an excessive function, bounded by 1 and $P_B p_n \leq p_n$. Therefore,

$$p_n - P_B p_n \in [0, 1].$$

Let us set $T_0 := 0$ and $T_1, T_2, T_3, \ldots$ to be times of successive visits to A, then to B, then back to A and so on. Formally, these times are defined as

$$T_1 := T_A,$$
$$T_2 := T_A + T_B \circ \theta_{T_A},$$
$$\vdots$$
$$T_{2n+1} := T_{2n} + T_A \circ \theta_{T_{2n}},$$
$$T_{2n+2} := T_{2n+1} + T_B \circ \theta_{T_{2n+1}},$$

where θ is the shift operator. An induction argument shows that

$$P_{T_{2n}} = (P_A P_B)^n, \quad n \in \mathbb{N}.$$

Then it can be easily checked that

$$\mathbb{P}_x[T_A < T_B, T_{2n+1} \leq L \leq T_{2n+2}] = p_n(x) - P_B p_n(x),$$

where L is the last exit time from A, i.e.,

$$L = L_A = \sup\{t > 0 | x_t \in A\}.$$

L is a.s. finite because usually we suppose that our process is transient, in the sense that if it enters a set then it must leave it also. $\square$

Theorem 10.3 *State-constrained reachability probability* p_A^B *solves the following boundary value problem*:

$$\begin{cases} \mathscr{L} p(x) = 0, & x \in \mathbf{X} \backslash (A \cup B) \\ p(x) = 1, & x \in A \\ p(x) = 0, & x \in B, \end{cases} \tag{10.2}$$

where $\mathscr{L}$ *is the infinitesimal generator of the given stochastic hybrid process.*

This is the main theorem about the characterisation of the state-constrained reachability. The theorem can be proved for Borel right processes that are stochastic hybrid processes. Stochastic hybrid processes have a continuous dynamics given by some diffusion processes and a discrete dynamics described by a Markov chain. Therefore, the proof is a consequence of the following two lemmas, which are instantiations of the theorem for Brownian motion and Markov chains. We have not found the proofs in any monograph of stochastic processes that treat first time passage problems, excursion theory for Markov processes—see for example [109], therefore we sketch these proofs in the following.

Lemma 10.4 *Let us consider a (discrete time, discrete state) Markov chain (X_t) with the state space Γ and the one-step transition function $p_1(x, y)$. Given two disjoint sets $A, B \subset \Gamma$, then the state-constrained reachability probability $p_A^B(x)$ is the solution of the boundary value problem*

$$
\begin{cases}
(1 - p_1)p(x) = 0, & x \in \Gamma \backslash (A \cup B) \\
p(x) = 1, & x \in A \\
p(x) = 0, & x \in B.
\end{cases}
\tag{10.3}
$$

Lemma 10.5 *For a discrete space Markov chain, it is known that its infinitesimal generator is given by*

$$
\mathscr{L} = 1 - p_1.
$$

Proof If $x \notin A \cup B$, we make the elementary remark that the first step away leads either to B and the event $\{T_A < T_B\}$ fails to happen, or to A, in which case the event happens, or to another point $y \notin A \cup B$, in which case the event happens with probability $P_y[T_A < T_B]$. Therefore, we obtain

$$
\mathbb{P}_x[T_A < T_B] = \sum_{y \in A} p_1(x, y) + \sum_{y \notin A \cup B} P_y[T_A < T_B].
$$

Then for $x \notin A \cup B$, we obtain

$$
p(x) = \sum_{y \in \Gamma} p_1(x, y)p(y).
$$

This ends the proof.   □

Lemma 10.6 *Let us consider W the standard d-dimensional Wiener process. Let A, B be two disjoint capacitable sets (see [51] for full definition) of nonzero capacity such that $A \cup B$ is closed. The reachability probability $p(x)$ satisfies the Laplace problem*

$$
\nabla^2 p(x) = 0
$$

on $\mathbf{X} - (A \cup B)$ with the boundary condition

$$
p(x) =
\begin{cases}
1, & if \ x \in A \\
0, & if \ x \in B.
\end{cases}
$$

Proof Let $x \in \mathbf{X} - (A \cup B)$ and H be a ball of radius h and surface S in $\mathbf{X} - (A \cup B)$ centred in x. Define the random variable

$$
T = \inf\{t \,|\, x_t(\omega) \in S\}.
$$

This has the property that

$$\mathbb{P}_x[T < \infty] = 1.$$

Let

$$H_i := \left\{ |W(i) - W(i-1)| \leq 2h \right\}.$$

Then $\mathbb{P}_x(A_1) < 1$ and $\lim_{n \to \infty} \mathbb{P}_x[T > n] = 0$. This results from the inequality

$$\mathbb{P}_x[T > n] \leq \mathbb{P}_x(A_1 \cap \cdots \cap A_n) = \mathbb{P}_x(A_1)^n.$$

We have

$$p(x) = \int_{y \in S} \mathbb{P}_x[T_A < T_B | W(T) = y] f(y) \, dS,$$

where $f(y) = 1/|S|$ is the density function. This means that

$$p(x) = \int_{y \in S} p(y)/|S| \, dS. \qquad \qquad \square$$

Theorem 10.3 characterises the probabilities of the state-constrained reachability as the solutions for a Dirichlet boundary value problem (DBVP) associated to the generator of the underlying stochastic hybrid process. This generator is a second order elliptic integro-differential operator and it is known that, for this type of non-local operator, the value of the solutions for the DBVP has to be prescribed not only on the boundary of the domain but also in its whole complementary set [95]. Under standard hypotheses, the existence and uniqueness of the solutions for such equations can be proved. The solutions are called *potential functions* for the underlying Markov process and they play the same role as harmonic functions for the Laplace operator.

Dirichlet boundary value problems for such operators already have been addressed in the literature by different theories:

- for a classical PDE approach using Sobolev spaces, see [95];
- for a probabilistic approach using the operator semigroup, see [222];
- for a viscosity solution approach, see [11].

For the verification problems defined in the context of stochastic hybrid systems, the DBVP defined in Theorem 10.3 will have to be solved only locally in the appropriate modes. In this way, the quite difficult boundary condition (that results from the definition of the infinitesimal generator) can be avoided. Then we consider that most numerical solutions available for the Dirichlet boundary value problem corresponding to second order elliptic partial differential operators can be extended in a satisfactory manner to solve our problem.

Without doubt, we need to consider the applications of state-constrained reachability in the framework of controlled stochastic hybrid systems. The control can be defined either

- at the 'low level', i.e., continuous dynamics in the operating modes are controlled diffusions,

- or, at the 'decision level', i.e., the jumping structure is governed by a decision process (usually in the form of a Markov decision process).

10.4 State-Constrained Reachability in a Development Setting

So far, state-constrained reachability has been defined and investigated in the abstract setting provided by the SHS models. We have shown that the reach probabilities represent the solution of an elliptic integro-differential equation. This problem can be efficiently solved in some cases, for example in the case where the equation can be reduced to an elliptic one. However, for complex systems, the integro-differential equation can be difficult to solve. In this case, the abstract model of SHS should be replaced with more manageable models. This can be done in many ways, like through approximations, functional abstractions, model reductions and so on. Based on existing research we propose a multilayer approach for describing a complex hybrid system with stochastic behaviour. The same system is described using a set of different models, each one constructed at a different level of abstraction. The models can be related by abstraction or refinement maps. This approach makes it possible for us to solve a specific problem for the given system at the right level of abstraction.

In this setting, we search for more manageable solutions for state-constrained reachability analysis.

10.4.1 Multilayer Stochastic Hybrid Systems

In this subsection, we introduce a multilayer model for stochastic hybrid systems. This is inspired by the *viewpoints model* from software engineering. There, a system is modularly developed from different perspectives. Each perspective provides a model of the system, called a viewpoint. Then the viewpoints need to be consistently unified to provide the overall system description. This methodology corresponds to a horizontal development philosophy. In this section, we proposed a vertical (or hierarchical) viewpoint approach. In this approach, the system is described by viewpoints constructed on top of each other, each one providing a partial model at a different abstraction level.

Mathematically, at the level j, a *viewpoint* is an SHS

$$H^j := \left(\left(Q^j, d^j, \mathcal{X}^j \right), b^j, \sigma^j, \text{Init}^j, \lambda^j, R^j \right)$$

and all its elements (discrete/continuous states, trajectories, jumping times, etc.) carry the superscript j.

At the level 0, the corresponding viewpoint is an SHS,

$$H^0 := \left(\left(Q^0, d^0, \mathcal{X}^0 \right), b^0, \sigma^0, \text{Init}^0, \lambda^0, R^0 \right).$$

The viewpoint H^j is related to viewpoint of level $(j-1)$,

$$H^{j-1} := \left(\left(Q^{j-1}, d^{j-1}, \mathscr{X}^{j-1}\right), b^{j-1}, \sigma^{j-1}, \text{Init}^{j-1}, \lambda^{j-1}, R^{j-1}\right),$$

by a pair of maps (Φ, Ψ), where

$$\Phi : X^{j-1} \to X^j$$

relates the states and

$$\Psi : \Omega^j \to \Omega^{j-1}$$

relates the trajectories. In relational algebra, such a pair is called a *twisted relation*.

The first map Φ is a surjective map that describes how H^j simulates H^{j-1} by means of the following property:

$$\mathscr{L}^{j-1}(f \circ \Phi) = \mathscr{L}^j f \circ \Phi, \quad \forall f \in \mathscr{B}^b\left(X^j\right), \tag{10.4}$$

where $\mathscr{L}^{j-1}$ and $\mathscr{L}^j$ represent the infinitesimal generators for H^{j-1} and, respectively, for H^j. Relation (10.4) can be given as well in terms of transition probabilities or operator semigroups as follows:

$$P_t^{j-1}(f \circ \Phi)(u) = \left(P_t^j f\right)(\Phi(u)), \quad \forall f \in \mathscr{B}^b\left(X^j\right), u \in X^{j-1}.$$

The second map Ψ can be defined in various ways, adding flexibility to the viewpoint modelling. For example, Ψ can be generically defined by replacing a certain discrete transition in the dynamics of the viewpoint j with a set of trajectories of the hybrid system described by the viewpoint $(j-1)$. In particular, a single discrete transition can be mapped into the trajectories of a continuous dynamical system. In this way, a viewpoint described by a discrete transition system can be related to another viewpoint described by a hybrid system.

To illustrate the viewpoint approach let us consider a flying aircraft. At the lowest level of detail, its dynamics can be accurately described by a switching diffusion process. In this model of SHS, the hybrid trajectories are continuous and are piecewise diffusion paths. At the most abstract level, the flight can be modelled as a probabilistic timed automaton that is a sort of discrete transition system. In this viewpoint, the aircraft lies in a state for a certain time and then with a given rate it makes a discrete transition. An intermediate viewpoint, between the continuous and discrete, is modelled by a stochastic hybrid system. In the intermediate viewpoint, a discrete transition from the discrete viewpoint is refined into a continuous mode and certain continuous paths from the continuous viewpoint are abstracted away into discrete transitions. Of course, in the stochastic setting there can be many subtle cases, like rates of discrete transitions that depend on the diffusion evolution in a mode.

The utility of a multilayer model consists of the possibility of solving categories of problems at different levels/viewpoints. For example, the stability problems are more efficiently studied in a continuous viewpoint (corresponding to the lowest abstraction level). A safety verification problem can be formally tackled in a discrete viewpoint (corresponding to the highest abstraction level). Many control problems can be suitably studied in the hybrid viewpoint (corresponding to an intermediate level of abstraction).

State-constrained reachability has been defined in a viewpoint corresponding to an intermediate discrete/continuous level of abstraction. Since in the previous sections it has been proved that the approach in this viewpoint can lead to problems with integro-differential operators that can be difficult to solve, it is worthy to try to study the problem at other levels of abstractions/viewpoints. In a discrete viewpoint, one can hope to use probabilistic model checking techniques. In a continuous viewpoint, the well-developed mathematical apparatus of diffusion processes is becoming available with specific benefits.

In order to make the state-constrained reachability approach practical, a further refinement of the mathematical model is necessary. This refinement takes into account the Euclidean space, in which processes evolve. We call such processes *spatial SHS*. An SHS is called spatial if some of its parameters form together a subspace of the Euclidean spaces $\mathbb{R}$, $\mathbb{R}^2$ or $\mathbb{R}^3$.

A multilayer model can be fruitfully conjoined with spatial models for designing or for improving control. Suppose that a system that is an n-dimensional SHS is a spatial process in a higher level of abstraction obtained by ignoring the nonspatial parameters. With this, one can obtain a viewpoint in which the system is modelled as a spatial Wiener process. The state-constrained stochastic reachability problem becomes more tractable in this viewpoint. If the reach probability is high and causes of this fact can be detected, then in the original viewpoint a control strategy can be considered that minimises the reach probability. For example, in the case of an air traffic control system, the spatial viewpoint can indicate that the collision probability becomes higher in areas of dense traffic with no coordination. Then a control strategy will design a pathway for the aircraft that avoids the high traffic density regions.

Continuous Viewpoint   The following two examples are inspired from [109].

Example 10.7 Let A be the sphere with radius ε and the centre at the origin of $\mathbb{R}^3$. Let us consider the continuous viewpoint of an SHS modelled as a three-dimensional Wiener process W with $W_0 = x_0 \notin A$. The problem is to compute the probability that W visits A. Let B be a sphere with the radius R and centre at the origin, where R is much larger than ε ($\varepsilon \ll R$). We have to look for a solution of Laplace's equation in spherical polar coordinates:

$$\frac{\partial}{\partial r}\left(r^2 \frac{\partial p}{\partial r}\right) + \frac{1}{\sin\theta}\frac{\partial}{\partial \theta}\left(\sin\theta \frac{\partial p}{\partial \theta}\right) + \frac{1}{\sin^2\phi}\frac{\partial^2 p}{\partial \phi^2} = 0, \tag{10.5}$$

subject to boundary conditions

$$p(x) = \begin{cases} 1, & \text{if } x \in A, \\ 0, & \text{if } x \in B. \end{cases}$$

Solutions for Eq. (10.5) with spherical symmetry have the form

$$p(x) = \frac{c_1}{r} + c_2 \quad \text{if } x = (r, \theta, \phi).$$

Using the boundary conditions, the following solution can be obtained [109]:

$$p_R(x) = \frac{r^{-1} - R^{-1}}{\varepsilon^{-1} - R^{-1}}.$$

Making $R \to \infty$, we get

$$p_R(x) \to \mathbb{P}(T_A < \infty) = \frac{\varepsilon}{r}, \quad r > \varepsilon.$$

Example 10.8 Let us consider a two-dimensional Wiener process W with $W_0 = x_0$, which can be thought of as another spatial viewpoint for an SHS. Again we use the polar coordinates (r, θ) and suppose that W evolves in the set

$$-\pi < -\alpha \leq \theta \leq \alpha < \pi T.$$

If A is the line $\theta = \alpha$ and B is the line $\theta = -\alpha$, we may ask for the probability that W reaches A before B. We consider now the planar Laplace equation in polar coordinates

$$\frac{1}{r} \frac{\partial}{\partial r} \left(r \frac{\partial p}{\partial r} \right) + \frac{1}{r^2} \frac{\partial^2 p}{\partial \theta} = 0$$

subject to the boundary conditions

$$p = 1 \quad \text{on } \theta = \alpha; \qquad p = 0 \quad \text{on } \theta = -\alpha.$$

It can be checked [109] that the function

$$p = \frac{\theta + \alpha}{2\alpha}, \quad -\alpha \leq \theta \leq \alpha,$$

is the required solution and so

$$p(x_1, x_2) = \frac{1}{2\alpha} \left(\alpha + \tan^{-1} \frac{x_2}{x_1} \right).$$

Discrete Viewpoint

Example 10.9 Let us consider a discrete time Markov chain that can be thought of as a discrete viewpoint for an SHS. In the literature, the number of transitions (time) required before the state will move from i to j for the first time is referred to as the first passage time. It is possible to calculate the average (or expected) number of transitions for the passage from state i to j. Let m_{ij} be the expected first passage time (number of transitions from state i to j). Moving from i to j in exact one transition has probability p_{ij}. If this were not the case then the state would change to k ($\neq j$). The probability of moving from i to k (for all $k \neq j$) would be the sum of all the probabilities p_{ik} for all k ($\neq j$), i.e., $\sum_{k \neq j} p_{ik}$. We now need to move from i to j. This may require many transitions and, based on the Markov property, its probability is given by $\sum_{k \neq j} p_{ik} m_{kj}$. Then

$$m_{ij} = p_{ij} + \sum_{k \neq j} p_{ik} + \sum_{k \neq j} p_{ik} m_{kj}$$

or, finally,

$$m_{ij} = 1 + \sum_{k \neq j} p_{ik} m_{kj}.$$

10.4.2 Electrostatic Analogy

It is known that the theory of Markov processes is intimately connected with mathematical physics (see [33]). The solutions of the DBVP from Theorem 10.3 can be characterised as certain potential functions defined on the underlying Markov process state space. These describe the 'probability distributions' of a charge that is distributed on the state space where the set A (the obstacle) produces a repulsive force and the set B (the target) yields an attractive force. In potential theory, the physical interpretation of state-constrained reachability probability considered in this chapter is related to the *condenser problem* (see [74]). This is described as follows: suppose there are given two disjoint compact conductors A, B in the Euclidean space $\mathbb{R}^3$ of positive capacity [51]. A positive electric unit charge placed on A and a negative unit charge on B, both allowed to distribute freely on the respective sets, will find a state of equilibrium, which is characterised on one hand by minimal energy and on the other hand by constant potential on A and on B (possibly taking out exceptional sets of zero capacity).

10.5 Further Generalisations: Randomised Targets

In this section, we investigate how to define the notion of the state-constrained reachability when the 'final goal' is replaced by a 'random event' that takes place at a random time. The objective is to compute the probability of those trajectories that visit a set before 'something' random happens.

Leveraging State-Constrained Reachability   State-constrained reachability analysis seeks to obtain estimations for $\mathbb{P}_x[T_A < T_B]$, i.e., to find the probability of visiting B after the process has visited A. As we have seen in previous sections, this problem can be characterised as a boundary value problem. In practice, it may happen that the sets A and B are not explicitly given. These sets can be characterised as:

Fig. 10.3 Moving target

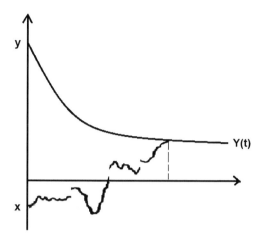

- level sets for some given functions,
- sets of states that validate some logical formulae,
- metastable sets for the given process.

Sometimes, for the computation of state-constrained reachability probabilities, more information is needed about at least one of these sets. Often, the available information is about the set B that can be either:

- the boundary of one mode of the stochastic hybrid process,
- a cemetery set where the process is trapped,
- a set that is reached according to a state-dependent rate and so on.

Moreover, in the expression of the state-constrained reachability probability, we may replace the hitting time T_B of B with a suitable random time that could be, for example:

- a discrete transition time from one mode to another of the stochastic hybrid process,
- a time of the apparition of a certain event,
- a time defined by the until operator in a suitable continuous stochastic logic associated to our hybrid Markov process.

Then the probabilities that should be computed are

$$\mathbb{P}_x[T_A < T],$$

where T is a random time that will be defined properly in the remainder of this section.

The idea of a moving target is illustrated in Fig. 10.3.

Randomised Stopping Times We enlarge the space of stopping times with the so-called randomised stopping times [15]. A *randomised stopping time* T is defined to be a map

$$T : \Omega \times [0, 1] \to [0, \infty]$$

such that T is a stopping time with respect to the σ-algebras $(\mathscr{F}_t \times \mathscr{B}_1)$, where $\mathscr{B}_1$ represents the Borel σ-algebra of the interval $[0, 1]$. It is required that for every $\omega \in \Omega$, $T(\omega, \cdot)$ is nondecreasing and left continuous on $[0, 1]$.

If T is a randomised stopping time, then $T(\cdot, a)$ is an ordinary stopping time. A randomised stopping time T can be characterised by a *stopping time measure* (ω-distribution) K induced by T. K is defined as the map

$$K : \Omega \times \mathscr{B}[0, \infty] \to [0, 1],$$

$$K\big(\omega, [0, t]\big) := \sup\big\{a : T(\omega, a) \leq t\big\}, \tag{10.6}$$

provided that $K(\omega, \cdot)$ is a measure on $\mathscr{B}[0, \infty]$. $K(\omega, \cdot)$ is a version of the conditional distribution of T given the entire trajectory ω.

Using the measure K, one can get back the randomised stopping time T by

$$T(\omega, a) = \inf\big\{t : K\big(\omega, [0, t]\big) \geq a\big\}. \tag{10.7}$$

Moreover, one can define a stopping time measure as an independent mathematical object as follows.

A map

$$K : \Omega \times \mathscr{B}[0, \infty] \to [0, 1]$$

is called *stopping time measure* if

(i) $K(\omega, \cdot)$ is a probability measure for each $\omega \in \Omega$;
(ii) $K(\cdot, [0, t])$ is $\mathscr{F}_t$-measurable for each t.

Then T can be defined by (10.7). Therefore, there exists a complete correspondence between the notions of stopping time measure and randomised stopping time.

Usually, the following notation is in use:

$$K_t := K\big((t, \infty]\big) = K\big(\cdot, (t, \infty]\big).$$

If $\mathbb{P}$ is the underlying probability on $(\Omega, \mathscr{F})$ and m_1 is the Lebesgue measure on the unit interval $[0, 1]$, then K_t is a version of the conditional probability of $\{T > t\}$ using the probability $\mathbb{P} \times m_1$ with respect to the σ-algebra $\mathscr{F}$.

Markovian Randomised Stopping Times   Working with randomised stopping times might be a difficult task, since the Markov property is still true only with respect to the nonrandomised stopping times (*strong Markov property*). To keep things simpler, we consider only the Markovian randomised stopping times. Examples of this kind of random time are:

• *Markov killing times*: T is a Markov killing time for M if under $\mathbb{P}_x$ the killed process $(x_t | 0 \leq t \leq T)$ is Markovian with the sub-Markovian semigroup $(\Gamma_t)_{t \geq 0}$:

$$\Gamma_t f(x) := \mathbb{P}_x\big[f(x_t) 1_{(t < T)}\big].$$

In addition, we assume that $\Gamma_t f$ is $\mathscr{B}$-measurable for all $t > 0$ and all positive $\mathscr{B}$-measurable f.

- *Terminal times*: An $\mathscr{F}_t$-stopping time $T : \Omega \to \mathbb{R}_+$ is called a terminal time if

$$T = t + T \circ \theta_t \tag{10.8}$$

identically on $[t < T]$.

Clearly, relation (10.8) expresses the memoryless property of a terminal time,

$$T(\omega) - t = T(\theta_t \omega),$$

i.e., the value of T on a path ω after time t has elapsed is equal to the value of T on the same path shifted/translated with the time t.

Examples of killing times are:

- the random time T_r with the stopping measure $K_t = e^{-rt} \mathbf{P}_t$;
- $\infty = \lim_{r \to 0} T_r$;
- the first entrance/last exit time of a suitable subset B of the state space $\mathbf{X}$;
- the random time obtained by killing the process at state-dependent rate $k(x_t)$.

A finite fixed time T is not a Markov killing time unless the process is set up as a space–time process and then T becomes a hitting time. As shown by the example of last exit times, the killing times may not be stopping times of the process, in comparison with the terminal times, which necessarily should be stopping times.

The most common examples of terminal times are provided by the hitting times of measurable subsets of state space $\mathbf{X}$. The jumping times in the definition of a stochastic hybrid process are terminal times.

Multiplicative Functionals   A common methodology to obtain randomised stopping times is by using *multiplicative functionals*. These functionals have a long history in the theory of Markov processes and they have been employed mostly to describe transformations of the trajectories for these processes.

The seminal work about properties of trajectories of a Markov process in connection with multiplicative functionals belongs to E.B. Dynkin [77]. Under some regularity hypotheses, Dynkin proved that transformed processes can be defined such that their trajectories are 'restrictions' of the trajectories of the initial process. In [154], Kunita and Watanabe showed that the transformations of a Markov process governed by multiplicative functionals, whose expectations are dominated by 1, preserve some regularity properties of the initial process. Such properties are: (strong) Markovianity, right continuity of the trajectories and quasi-left continuity of the stopping time sequences.

A systematic study of the multiplicative functionals of Markov processes was done in [33] and later in [208]. A *stopping time measure* α will be called *multiplicative functional* if for every $s, t \geq 0$,

$$\alpha_{t+s} = \alpha_t (\alpha_s \circ \theta_t) \quad \text{a.s.} \tag{10.9}$$

We assume that α is an *exact* multiplicative functional [33]. The property (10.9) ensures that the randomised stopping time generated by (α_t) has the memoryless property.

The study of stochastic reachability when there are constraints regarding:

- the target goal, which may have its own dynamics (as a set or a point mass), or may represent the boundary of a safe set,
- the time horizon that could be a randomised stopping time,

represents ongoing research. Partial answers of this problem will be given in a subsequent volume to this book.

10.6 Some Remarks

In this chapter, we extended the so-called constrained reachability problem from the probabilistic discrete case to stochastic hybrid systems. Then we mathematically defined this problem and we obtained the reach probabilities as solutions of a boundary value problem. These characterisations are useful in stochastic control and in probabilistic path planning. In this chapter, the stochastic reachability problem for stochastic hybrid systems has been specialised by introducing constraints relative to the state space. We proved that state-constrained stochastic reachability is solvable. Moreover, we described how the concept could be leveraged so that it could capture more constraints regarding time and space. Numerical solutions for this problem depend fairly on the underlying model of stochastic hybrid systems. It is clear that the simplest way to deal with this problem is to work either at the bottom level (using only diffusions) or the higher level (using only Markov chains).

Chapter 11
Applications of Stochastic Reachability

11.1 Overview

This chapter aims to give a flavour of the range of stochastic reachability applications. A list with appropriate references will be given. The list is not exhaustive; the interested reader may search for more exciting applications of this concept in different areas. Here, we point out only the applications in air traffic management and biology.

Refinements of the notion of stochastic reachability have led to a huge research palette of applications that are difficult to describe in only a few lines. Therefore, we stick only to the 'classical' application domains, which have driven the extraordinary growing body of research in the area of stochastic reachability.

It is very important to mention here that the study of stochastic reachability has impelled new cross- and interdisciplinary research topics like:

- modelling paradigms that can be studied from various perspectives (mathematics, control theory, and computer science);
- hidden aspects in the connections between concepts (hitting probabilities, optimal stopping, Choquet capacity);
- possibility of integration of the existing methods in different disciplines for solving analysis, verification and control problems related to complex systems like the stochastic hybrid systems.

11.2 Applications of Stochastic Reachability in Air Traffic Management

We intend to describe the state of art in applying stochastic reachability analysis to solve problems from air traffic management (ATM). This section is motivated by the fact that stochastic reachability has emerged as a useful tool to solving problems that appear in safety critical situations in ATM.

L.M. Bujorianu, *Stochastic Reachability Analysis of Hybrid Systems*,
Communications and Control Engineering,
DOI 10.1007/978-1-4471-2795-6_11, © Springer-Verlag London Limited 2012

In air traffic, we deal with an event of mid-air collision at the moment when two aircraft strike each other. Such an event can be thought of as the instant when the joint state of the aircraft involved is visiting a certain subset of their joint state space. With this representation in mind, the problem of estimating the probability of collision between two aircraft within a finite horizon time is to analyse the probability of reaching the collision subset by the joint aircraft state. Such a problem has been defined in this monograph as the stochastic reachability problem.

In the last few decades, the demand for air travel has seen a huge growth and this raises great challenges to the current, mostly human operated ATM system, where air traffic controllers (ATCs) are responsible for air travel safety. The aircraft conflicts represent situations where an aircraft comes too close to another aircraft or enters a forbidden zone. One way to deal with safety issues is the introduction of a methodology that predicts aircraft conflicts and could reduce ATC workload, increasing in this way capacity without compromising safety.

Conflict Detection   The problem of the collision risk assessment has become challenging in free flight air traffic, where there is no limitation of a fixed route structure. This problem is known in the ATM community as the probabilistic conflict detection (CD) problem.

'Conflict detection is the process of identifying when a pair of aircraft in flight will enter some procedurally specified mutual exclusion zone, with sufficient warning that an alert issued to air traffic control will result in a straightforward and economic resolution of the problem' [226].

Conflict Resolution   During flight, aircraft have to maintain safety. In terms of air traffic control, this means that aircraft have to maintain a minimum separation between each other during the entire flight. Any violation of this minimum separation is defined as a 'conflict'. The object of conflict resolution (CR) is to detect impending conflicts and resolve them by changing flight plans, subject to constraints and priorities.

Solutions for conflict resolution problems can be divided into two main categories:

(i) *Centralised solutions*: In this approach, the CD and CR algorithm has to detect and resolve all impending conflicts in a specific airspace sector of interest in a centralised manner (i.e., on the ground).

(ii) *Decentralised solutions*: This approach is known as free flight or self separation [30]. In this case, the aircraft themselves are responsible for maintaining separation, detecting and resolving all impending conflicts in the airspace. This approach, even though it is more complicated and seems to be futuristic, is believed to indeed be the solution for the increased air transportation, taking advantage of the more accurate information that aircraft have onboard compared to what ATCs have on the ground.

11.2.1 Work of Blom and Co-workers

Dr. Henk A.P. Blom is senior researcher at the National Laboratory of Aerospace in Amsterdam, the Netherlands. In the last ten years, Dr. Blom has been the coordinator of two very large European Projects in Air Traffic Management: Hybridge [126] and iFLY [133]. He is one of the pioneers of the free flight concept [30–32]. He is a promoter of using stochastic hybrid systems for modelling the dynamics of aircraft [32] and stochastic reachability for collision risk analysis [31]. Moreover, he is the first one to use general stochastic hybrid systems and their connection with dynamically coloured Petri nets for modelling aircraft dynamics [84–86]. In [30], Blom and co-workers start with simple Monte Carlo (MC) simulations for mid-air collision probability. The idea is to run many MC simulations with a free flight stochastic hybrid model and count the fractions of runs for which a collision occurs. The main advantage of MC simulations is that they do not require peculiar assumptions regarding the underlying stochastic model. The main problem is that in order to obtain accurate estimations of reach probabilities, huge simulation time is required. To tackle this problem, the above approach considered the development of a sequential MC simulation technique for estimating small reach probabilities, which includes a characterisation of convergence behaviour. The underlying idea is to express the small probabilities that have to be estimated as the product of a certain number of larger probabilities that are estimated by the MC method. The reach probability is obtained by introducing a sequence of intermediate sets that are visited in an ordered sequential way, before the final goal states are reached. The desired reach probability is obtained as the product of conditional probabilities of visiting a set of intermediate states provided that the previous set of intermediate states has been reached. The estimation of each conditional probability is obtained by simulating in parallel several copies of the system (each copy is thought of as a particle whose trajectory is governed by the system dynamics). The system model must satisfy the strong Markov property in order to ensure unbiased estimations of reach probabilities.

11.2.2 Work of Lygeros and Co-workers

Professor John Lygeros is the Head of the Control Group at ETH, Zurich. He is a promoter of (deterministic and stochastic) hybrid systems and the corresponding reachability analysis in the modelling and safety certification of the air traffic management system. His group's recent work in the area of conflict detection and resolution and trajectory prediction has to be mentioned here. Some references (to record only a few) regarding these problems are [172–175].

First, the main achievement of this work was to develop a complete model and a fast-time simulator for the validation of trajectory prediction, conflict detection and conflict resolution algorithms. The model allows simulation of multiple flights taking place at the same time. The principal characteristics of the model are stochasticity and hybridicity. The group's goal is to produce an efficient automated conflict resolution algorithm based on different cost criteria.

The technique proposed by Lygeros's group, widely used in robotics, which can guarantee collision avoidance between the moving aircraft, is based on the navigation function/potential field methods. This method has been used with great success in several robotics applications. Its use in the air traffic control framework is much more complicated since there exist additional dynamic constraints (e.g., speed constraints) for the aircraft. Applying directly this method for CD and CR produces conflict-free trajectories, but the operational constraints are not considered. To surpass this drawback, the group has proposed the use of model predictive control, which is a control technique able to deal with system constraints. The combination of the two techniques led to another difficulty, the intractability of the control scheme. To overcome this difficulty, they use randomised optimisation techniques, which can deal with complex systems under certain appropriate assumptions. The methods are based on a variation of simulated annealing, using Markov chain Monte Carlo simulations.

11.2.3 Work of Prandini and Hu

Both Professors Maria Prandini (Milan University) and Jianghai Hu (Purdue University) have been working for a long time in stochastic modelling and verification of the air traffic management system [124, 125, 127, 197]. The main achievement of their work is the fact that they use stochastic reachability analysis defined in the stochastic hybrid system framework to provide solutions for aircraft conflict detection.

In [127, 197], Prandini and Hu formulate the problem of aircraft conflict detection as a stochastic reachability analysis problem for a switching diffusion process. They use a grid-based computation to estimate the probability that two aircraft come closer to each other than some established minimum separation criterion. Simulations are also reported to show the efficiency of the approach. The drawback of this approach is the fact that the approximations of switching diffusion processes lead to Markov chains with very large state spaces, in which event the proposed numerical methods might be very slow.

11.3 Applications of Stochastic Reachability in Biology

A very nice application of stochastic reachability for the analysis of complex cellular networks has been reported in [80]. Cellular networks implementing life's programs operate in a random cell environment and they are influenced by different kinds of noise (biochemical, thermal) from the environment. Rare events produced by the dynamics of noise in such networks appear in the form of some large deviations that can lead to important failures of the cellular machinery, resulting in cancer or releasing viruses from their quiescent state. The above reference provides upper bounds for the probability of such rare events based on the barrier certification method.

Modelling, analysis and verification methods of biochemical systems in the SHS framework have been reported in [201]. Case studies include models of sugar cataract development in the lens of a human eye [203], biodiesel production system [202], glycolysis that is a cellular energy conversion mechanism found in every living cell [204], and so on. All these case studies will be expanded in a subsequent volume of this book.

Modelling for gene regulatory networks using stochastic hybrid systems has been reported in [123]. In *Bacillus subtilis*, the genetic network regulating the biosynthesis of subtilin is modelled as a stochastic hybrid system. The continuous state of the hybrid system describes concentrations of subtilin and various regulating proteins, whose productions are controlled by switches modelled as Markov chains. Analysis and simulation results show that the stochastic model is more appropriate for explaining the intrinsic random fluctuations in gene expressions when there exists only a small number of participating players (i.e., low concentrations of regulatory proteins intracellularly and a small cell population extracellularly).

The case studies existing in biology modelled using stochastic hybrid systems show that the analysis of such systems should include not only reachability problems, but also ergodicity problems (long time behaviours) and invariants (invariant measures, invariant sets).

11.4 Some Remarks

The research area for stochastic reachability is a continuously growing field of stochastic hybrid control. Social problems of society like free flight safety and smart grid reliability can find appropriate study frameworks in the context of stochastic hybrid systems and stochastic reachability. In this chapter, we named only a few application areas whose scientists have contributed to the flourishing of the stochastic reachability research field.

Appendix

A.1 Ordinary Differential Equations

An ordinary differential equation (or ODE) is an equation involving derivatives of an unknown quantity with respect to a single variable. An nth order ordinary differential equation is an equation of the form

$$F\left(t, x(t), \dot{x}(t), \ddot{x}(t), x^{(3)}(t), \ldots, x^{(n)}(t)\right) = 0 \tag{A.1}$$

where $F : \mathrm{Dom}(F) \subseteq \mathbb{R} \times E \times E \times \cdots \times E \to \mathbb{R}^j$.

If $I \subseteq \mathbb{R}$ is an interval, then $x : I \to E$ is said to be a solution of (A.1) on I if x has derivatives up to order n at every $t \in I$ and those derivatives satisfy (A.1). Often, the dependence of x on t is suppressed. Also, there will often be side conditions given that narrow down the set of solutions. In this class, we will concentrate on initial conditions that prescribe $x^{(l)}(t_0)$ for some initial time $t_0 \in \mathbb{R}$ and some choices of $l \in \{0, 1, \ldots, n\}$. Some ODE classes study two-point boundary value problems, in which the value of a function and its derivatives at two different points are required to satisfy given algebraic equations, but we do not present them here.

Every ODE can be transformed into an equivalent first order equation. Consider the first order initial-value problem (or IVP):

$$\begin{cases} F(t, x(t), \dot{x}(t)) = 0 \\ x(t_0) = x_0 \\ \dot{x}(t_0) = p_0, \end{cases}$$

where $F : \mathrm{dom}(F) \subset \mathbb{R} \times \mathbb{R}^n \times \mathbb{R}^n \to \mathbb{R}^n$. Using the implicit function theorem, one can prove that this equation is equivalent to

$$\begin{cases} \dot{x} = f(t, x) \\ x(t_0) = x_0 \end{cases} \tag{A.2}$$

for an appropriate $f : \mathrm{dom}(f) \subset \mathbb{R} \times \mathbb{R}^n \to \mathbb{R}^n$ with (t_0, x_0) in the interior of $\mathrm{dom}(f)$. The ODE (A.2) is autonomous if f does not really depend on t, i.e., if $\mathrm{dom}(f) = \mathbb{R} \times \Omega$ for some $\Omega \subset \mathbb{R}^n$ and there is a function $g : \Omega \to \mathbb{R}^n$ such that $f(t, u) = g(u)$ for every $t \in \mathbb{R}$ and every $u \in \Omega$. It can be proved that every nonautonomous ODE is actually equivalent to an autonomous ODE.

L.M. Bujorianu, *Stochastic Reachability Analysis of Hybrid Systems*,
Communications and Control Engineering,
DOI 10.1007/978-1-4471-2795-6, © Springer-Verlag London Limited 2012

Theorem A.1 (Existence and uniqueness of solutions) *Suppose $f \in C(U, \mathbb{R}^n)$, where U is an open subset of $\mathbb{R}^{n+1}$ and $(t_0, x_0) \in U$. If f is locally Lipschitz continuous in the second argument, uniformly with respect to the first, then there exists a unique local solution $x(t) \in C^1(I)$ of the IVP (A.2), where I is some interval around t_0.*

The hypothesis that f is locally Lipschitz continuous in the second argument, uniformly with respect to the first argument, means that, for every compact set $V_0 \subset U$, the following number:

$$L = \sup_{(t,x)\neq(t,y)\in V_0} \frac{|f(t, x) - f(t, y)|}{|x - y|},$$

which depends on V_0, is finite.

A.2 Dynamical Systems

A dynamical system, roughly speaking, is a set of objects which are allowed to change over time, whilst obeying a set of rules. One can think of a dynamical system as the time evolution of some physical system, such as the motion of a satellite planet under the influence of its respective gravitational forces or the evolution of the number of particles in a biochemical reaction.

Examples of dynamical systems include chemical plants, population growth of a species and the behaviour of a nation's economic structure. Many complex dynamical systems have to be systematically analysed. A well-developed theory of dynamical systems is available in literature (see, for example, [8, 215]). Systems in this class are associated, in one way or another, with algebraic, difference or differential equations, which are used to represent the behaviour of the dynamical system.

In engineering and mathematics, a dynamical system is a deterministic process, in which a function's value changes over time according to a rule that is defined in terms of the function's current value. Generally, dynamical systems come in two flavours: discrete and continuous. A discrete dynamical system involves step-by-step state changes. A dynamical system is called 'discrete' if time is measured in discrete steps; these are modelled as recursive relations. If time is measured continuously, the resulting continuous dynamical systems are expressed as ordinary differential equations. In contrast, a discrete dynamical system's behaviour is described using a transition relation or state space graph. For instance, a finite automaton is a discrete dynamical system.

A dynamical system is a semigroup T acting on a space X. That is, there is a map

$$\begin{cases} \phi : T \times X \to X \\ (t, x) \mapsto \phi_t(x) \end{cases}$$

such that

$$\phi_t \diamond \phi_s = \phi_{t+s}.$$

If T is a group, we will speak of an invertible dynamical system. Usually, T denotes time. We are mainly interested in *discrete dynamical systems* where $T = \mathbb{N}$ or $T = \mathbb{Z}$, or *continuous dynamical systems* where $T = \mathbb{R}_+$ or $T = \mathbb{R}$.

The prototypical example of a continuous dynamical system is the flow of an autonomous differential equation. Starting with the ODE framework, by placing conditions on the function $f : \Omega \subset \mathbb{R}^n \to \mathbb{R}^n$ and the point $x_0 \in \Omega$ we can guarantee that the autonomous IVP

$$\begin{cases} \dot{x} = f(x) \\ x(0) = x_0 \end{cases} \tag{A.3}$$

has a solution defined on some interval I containing 0 in its interior and this solution will be unique (up to restriction or extension). The function $f : X \to \mathbb{R}^n$ is called the *vector field* on $\mathbb{R}^n$. Moreover, it is possible to 'splice' together solutions of (A.3) in a natural way and, in fact, obtain solutions to IVPs with different initial times. These considerations lead us to study the concept of dynamical system.

In particular, solutions of the IVP (A.3) are also called *integral curves* or *trajectories*. We will say that a solution is an integral curve at x_0 if it satisfies $\phi(0) = x_0$.

There is a (unique) maximal integral curve ϕ_x at every point x, defined on a maximal interval $I_x = (T\_(x), T_+(x))$. Let

$$W := \bigcup_{x \in X} I_x \times \{x\}.$$

Then we define the *flow* of our differential equation to be the map

$$\begin{cases} \phi : W \to X \\ (t, x) \mapsto \phi(t, x), \end{cases}$$

where $\phi(t, x)$ is the maximal integral curve at x. Sometimes, we use the notation $\phi_t(x) = \phi_x(t) = \phi(t, x)$. It can be proved that the flow of an ODE satisfies the definition of a dynamical system.

A *plant* or an ODE with *inputs and outputs* is given by

$$\dot{x}(t) = f(x(t), u(t)),$$
$$y(t) = g(x(t)),$$

where the number of components of the state vector, n, is called the order of the system. The input u and the output y have p and q components, respectively, i.e., $u(t) \in U \subset \mathbb{R}^p$, $y(t) \subset \mathbb{R}^q$, $f : \mathbb{R}^n \times \mathbb{R}^p \to \mathbb{R}^n$ and $g : \mathbb{R}^n \to \mathbb{R}^q$.

A.3 Probabilistic Measure Theory

The main tool for studying stochastic processes is probability theory. The modern approach on probability theory is provided by the Kolmogorov axiomatic development.

Measurable Space   A measurable space $(\Omega, \mathcal{F})$ is a pair consisting of a *sample space* Ω together with a σ-algebra $\mathcal{F}$ of subsets of Ω. A σ-*algebra* or a σ-field $\mathcal{F}$ is a collection of subsets of Ω such that

1. $\Omega \in \mathcal{F}$ (the entire sample space belongs to the σ-algebra);
2. if $A \in \mathcal{F}$, then $A^c = \{\omega : \omega \notin A\} \in \mathcal{F}$ (σ-algebra is closed to the complementation operation);
3. if $\{A_i\}_{i\in\mathbb{N}} \subset \mathcal{F}$ then $\bigcup_{i\in\mathbb{N}} \in \mathcal{F}$ (σ-algebra is closed to the countable union operation).

From de Morgan's law of elementary set theory, it follows that

$$\left(\bigcap_{i\in\mathbb{N}} A_i\right)^c = \left(\bigcup_{i\in\mathbb{N}} A_i^c\right) \in \mathcal{F}.$$

A σ-algebra is called also an *event space* is a collection of subsets (called *events* or *measurable sets*) of a sample space such that any countable sequence of set theoretic operations (union, intersection, complementation) on events produces other events. The entire space Ω and the trivial event $\emptyset = \Omega^c$ are always in the event space.

A collection $\mathcal{M}$ of subsets of Ω is called a monotone class if it has the property that if $F_n \in \mathcal{M}$ for $n = 1, 2, \ldots$ and either $F_n \uparrow F$ (i.e., $F_n \subset F_{n+1}$ and $F = \bigcup_{n=1}^{\infty}$) or $F_n \downarrow F$ (i.e., $F_n \supset F_{n+1}$ and $F = \bigcap_{n=1}^{\infty}$) then also $F \in \mathcal{M}$. Clearly, any σ-algebra is a monotone class, but the reverse need not be true.

If (X, τ) is a topological space then the *Borel σ-algebra* $\mathcal{B}(X)$ in X is the σ-algebra generated by the open sets of τ, i.e., the smallest σ-algebra that contains all open sets.

Analytic Sets   A *paving* of a set F is any family of subsets of F that contains the empty set. The pair $(F, \mathcal{E})$ consisting of a set F and a paving $\mathcal{E}$ is called a *paved set*. The closure of a family of subsets $\mathcal{E}$ under countable unions (respectively, intersections) is denoted by $\mathcal{E}_\sigma$ (respectively, $\mathcal{E}_\delta$). We shall write $\mathcal{E}_{\sigma\delta} = (\mathcal{E}_\sigma)_\delta$.

Let $(F, \mathcal{E})$ be a paved set. A subset $A \in \mathcal{E}$ is called $\mathcal{E}$-*analytic* if there exists an auxiliary compact metrisable space $\mathbf{K}$ and a subset $B \subseteq K \times F$ belonging to $(\mathcal{K}(\mathbf{K}) \times \mathcal{E})_{\sigma\delta}$, such that A is the projection of B onto F. The paving of F consisting of all $\mathcal{E}$-analytic sets is denoted by $\mathcal{A}(\mathcal{E})$.

Remark A.2 $\mathcal{E} \subset \mathcal{A}(\mathcal{E})$ and the paving $\mathcal{A}(\mathcal{E})$ is closed under countable unions and intersections.

Let $\mathcal{B}(\mathbb{R})$ be the Borel sets in $\mathbb{R}$, $\mathcal{K}(\mathbb{R})$ the paving of all compact sets in $\mathbb{R}$ and $(\Omega, \mathcal{E})$ be a measurable space.

Theorem A.3 [22]

1. $\mathcal{B}(\mathbb{R}) \subset \mathcal{A}(\mathcal{K})$, $\mathcal{A}(\mathcal{B}(\mathbb{R})) = \mathcal{A}(\mathcal{K})$.
2. *The product σ-field $\mathcal{G} = \mathcal{B}(\mathbb{R}) \times \mathcal{E}$ on $\mathbb{R} \times \Omega$ is contained in $\mathcal{A}(\mathcal{K}(\mathbb{R}) \times \mathcal{E})$.*

3. *The projection onto Ω of an element of $\mathcal{G}$ (or, more generally, of $\mathcal{A}(\mathcal{G})$) is $\mathcal{E}$-analytic.*

Recall that a *Borel space* is a topological space that is *homeomorphic to a Borel subset of a complete separable metric space.*

Proposition A.4 [22] *Every Borel subset of a Borel space is analytic.*

Proposition A.5 [22] *Every analytic subset of a Borel space is universally measurable.*

Probability Space   A *probability space* $(\Omega, \mathcal{F}, \mathbb{P})$ is a triple consisting of a sample space Ω, a σ-algebra $\mathcal{F}$ of subsets of Ω and a probability measure $\mathbb{P}$ defined on the σ-algebra $\mathcal{F}$, i.e., $\mathbb{P}$ assigns a real number to every event $A \in \mathcal{F}$ such that the following axioms hold:

1. Non-negativity:

$$\mathbb{P}(A) \geq 0, \quad \forall A \in \mathcal{F}.$$

2. Normalisation:

$$\mathbb{P}(\Omega) = 1.$$

3. Countable additivity:

$$\mathbb{P}\left(\bigcup_{i=1}^{\infty} A\right) = \sum_{i=1}^{\infty} \mathbb{P}(A_i)$$

provided that $\{A_i\}_{i=1}^{\infty} \subset \mathcal{F}$ and $A_i \cap A_j = \emptyset$ for all $i \neq j$.

If the normalisation axiom does not hold, then $\mathbb{P}$ is simply called a *measure* and the triple $(\Omega, \mathcal{F}, \mathbb{P})$ is called a *measure space*.

In particular, we have

$$\mathbb{P}(\emptyset) = 0.$$

Of course, there may exist other sets $N \in \mathcal{F}$ such that $\mathbb{P}(N) = 0$. These are *sets of probability zero* or *null sets*. Because a null set N is interpreted as an event that occurs with probability zero, then any subset of N ought to be a null set. But there is no guarantee that such sets are indeed measurable, so no probability may be attached to them. In this case, we need the completion of $\mathcal{F}$ with respect to the probability $\mathbb{P}$ defined as follows:

$$\mathcal{F}^{\mathbb{P}} := \{F \cup N_1 : F \in \mathcal{F} \text{ and } N_1 \subset N \in \mathcal{F} \text{ with } \mathbb{P}(N) = 0\}.$$

Naturally, we define $\mathbb{P}(F \cup N_1) = \mathbb{P}(F)$. Then $\mathcal{F}^{\mathbb{P}}$ is a σ-algebra that includes $\mathcal{F}$ and is called the $\mathbb{P}$-*completion of* $\mathcal{F}$.

Defining a probability measure involves a rather difficult task: to give its value for all measurable sets. However, it might be possible to define a probability on a much smaller class of sets and then to extend it to all the remaining sets.

A class C is called a *field* if

1. $A \in C$ implies $A^c \in C$; and
2. $A_1, A_2 \in C$ implies $A_1 \cup A_2 \in C$.

The *Carathéodory extension theorem* states that if $\mathbb{P}$ is a countable additive set function on C with $\mathbb{P}(\Omega) = 1$ then there is a unique probability measure denoted by $\overline{\mathbb{P}}$ on $\sigma(C)$ (the smallest σ-algebra that contains C) such that $\overline{\mathbb{P}}(A) = \mathbb{P}(A)$ for all $A \in C$.

A.3.1 Random Variables

Given a measurable space $(\Omega, \mathcal{F})$, let $(\Gamma, \mathcal{G})$ be another measurable space. The first space can be thought of as an *input space* and the second as an *output space*. A random variable or measurable function defined on $(\Omega, \mathcal{F})$ and taking values in $(\Gamma, \mathcal{G})$ is a mapping or function

$$f : \Omega \to \Gamma$$

that satisfies the following property:

$$\text{if } B \in \mathcal{G}, \text{ then } f^{-1}(B) = \{\omega \in \Omega : f(\omega) \in B\} \in \mathcal{F}.$$

If f is a measurable function from $(\Omega, \mathcal{F})$ to $(\Gamma, \mathcal{G})$, we will say that f is $\mathcal{F}/\mathcal{G}$-measurable if the σ-algebras are not clear from the context.

If E is a subset of Ω, we define its *indicator function* as

$$\mathbf{1}_E(\omega) = \begin{cases} 1, & \omega \in E \\ 0, & \omega \notin E. \end{cases}$$

The indicator function $\mathbf{1}_E$ is a random variable if and only if E is a measurable set, i.e., $E \in \mathcal{F}$.

The name 'random variable' is commonly associated with the case when Γ is the real line and $\mathcal{G}$ is the σ-algebra generated by the topology of the real line. Other names used in the literature are *random vector* (when Γ is a Euclidean space) and *random process* (when Γ is a space of trajectories). We will use the term of *random variable* in the general sense.

A random variable is just a function with the property that the inverse images of the output events determined by the random variable are indeed events in the original measurable space. This property allows us to induce a probability measure for the output events as follows:

$$\mathbb{P}_f(B) = \mathbb{P}(f^{-1}(B)) = \mathbb{P}(f \in B), \quad B \in \mathcal{G}.$$

In this way, $(\Gamma, \mathcal{G}, \mathbb{P}_f)$ becomes a probability space since, due to the measurability of f, $\mathbb{P}_f$ is indeed a probability measure on $\mathcal{G}$. This induced probability measure is called the *distribution* or *law* of the random variable f. Let us denote by

$$\mathcal{F}_f = f^{-1}(\mathcal{G}) = \{f^{-1}(B) : B \in \mathcal{G}\}$$

the σ-field generated by f.

Theorem A.6 *Suppose that $g : \Omega \to \mathbb{R}$ is $\mathcal{F}_f/\mathcal{B}(\mathbb{R})$-measurable. Then there exists a measurable function $h : \Gamma \to \mathbb{R}$ such that*

$$g = h \circ f.$$

Therefore, g is measurable with respect to $\mathcal{F}_f$ if and only if g is a function of f.

If we have a collection of mappings $X := \{f_i : \Omega \to \Gamma\}$, then we denote by $\sigma(X)$ the smallest σ-algebra on Ω such that all the f_i become measurable.

Real Random Variables   Suppose that $X : \Omega \to \mathbb{R}$ is a 'traditional' random variable. Along with the distribution of X, we introduce its *distribution function*, usually denoted by F (or F_X, or F^X). By definition it is the function $F : \mathbb{R} \to [0, 1]$ defined by

$$F(x) = \mathbb{P}(X \le x).$$

Proposition A.7 *The distribution function of a random variable is right continuous, nondecreasing and satisfies $\lim_{x \to \infty} F(x) = 1$ and $\lim_{x \to -\infty} F(x) = 0$. The set of points where F is discontinuous is at most countable.*

For a random variable X, its distribution, i.e., the collection of all probabilities $\mathbb{P}(X \in B)$ with $B \in \mathcal{B}$, is determined by the distribution function F_X. Any function F on $\mathbb{R}$ that has the properties from Proposition A.7 is called a *distribution function*. The justification of the terminology lies in the fact that for any distribution function F, it is possible to construct a random variable on some probability space such that its distribution function is equal to F.

Theorem A.8 (Skorokhod's representation of a random variable with a given distribution function) *Let F be a distribution function on $\mathbb{R}$. Then there exists a probability space and a random variable $X : \Omega \to \mathbb{R}$ such that F is the distribution function of X.*

Gaussian Distribution   For the case when the state space is $\mathbb{R}$, we say that a Borel probability measure μ on $\mathbb{R}$ is a *Gaussian law* (or *Gaussian or normal distribution*) $N(a, \sigma^2)$, for $a \in \mathbb{R}$, $\sigma^2 > 0$, if its density with respect to the Lebesgue measure exists and is given by

$$f(x) = \frac{1}{(2\pi\sigma^2)^{1/2}} \exp\left[-\frac{1}{2\sigma^2}(x - a)^2\right], \quad x \in \mathbb{R}.$$

If $\sigma^2 = 0$ we set

$$N(a, 0) = \delta_a,$$

where δ_a is the Dirac measure in a. If $a = 0$, then $N(0, \sigma^2)$ is called a *centred Gaussian measure*.

Let $X : \Omega \to \mathbb{R}$ be a random variable. We say that X is a Gaussian random variable if its law μ is a Gaussian measure.

In particular, if $\mu = N(a, \sigma^2)$ we write

$$X \sim N(a, \sigma^2).$$

In this case, X has expectation

$$\mathbf{E}[X] = a$$

and variance

$$\mathrm{Var}(X) = \sigma^2.$$

Independence   Given a probability space $(\Omega, \mathcal{F}, \mathbb{P})$, two events $E, F \in \mathcal{F}$ are called *independent* if the product rule

$$\mathbb{P}(E \cap F) = \mathbb{P}(E)\mathbb{P}(F)$$

holds. This independence can be generalised to the independence of a sequence of events and to the independence of a sequence of σ-algebras.

We have the following definitions:

1. A sequence of σ-algebras $\mathcal{F}_1, \mathcal{F}_2, \ldots$ is called independent if for every $n \in \mathbb{N}$, we have

$$\mathbb{P}(E_1 \cap E_2 \cap \cdots \cap E_n) = \prod_{i=1}^{n} \mathbb{P}(E_i)$$

   for any choice of $E_i \in \mathcal{F}_i$, $i = 1, \ldots, n$.
2. A sequence of random variables $X_1, X_2, \ldots$ is called independent if the sequence of the σ-algebras $\sigma(X_1), \sigma(X_2), \ldots$ is independent.
3. A sequence of events $E_1, E_2, \ldots$ is called independent if the sequence of random variables $1_{E_1}, 1_{E_2}, \ldots$ is independent.

The above definitions also apply to finite sequences (of σ-algebras, random variables or events).

There exists a powerful result concerning sequences of events that is known as the Borel–Cantelli lemma. If $A_n \in \mathcal{F}$ for $n = 1, 2, \ldots$, we define

$$\limsup A_n = \bigcap_{n=1}^{\infty} \bigcup_{k=n}^{\infty} A_k, \qquad \liminf A_n = \bigcup_{n=1}^{\infty} \bigcap_{k=n}^{\infty} A_k.$$

Lemma A.9 (Borel–Cantelli) *Let $E_1, E_2, \ldots$ be a sequence of events.*

(i) *If it has the property that $\sum_{n=1}^{\infty} \mathbb{P}(E_n) < \infty$ then*

$$\mathbb{P}(\limsup E_n) = 0.$$

(ii) *If it has the property that $\sum_{n=1}^{\infty} \mathbb{P}(E_n) = \infty$ and, moreover, the sequence is independent, then*

$$\mathbb{P}(\limsup E_n) = 1.$$

Expectation and Integral   In elementary probability theory courses, usually we make a distinction between random variables X having a discrete distribution, for example on $\mathbb{N}$, and those having a density. In the former case, the expectation $\mathbb{E}X$ has expression $\sum_{k=1}^{\infty} k\mathbb{P}(X = k)$, whereas in the latter case it is $\mathbb{E}X = \int x f(x) \, dx$.

This distinction is not satisfactory from a mathematical point of view. Moreover, there exist random variables whose distributions are neither discrete, nor do they admit a density. These methods for computing expectations can be put under the same umbrella, if we consider expectations as special cases of the unifying concept of Lebesgue integral, which is a sophisticated way of integration that generalises the Riemann integral.

A random variable X is called *simple* if it can be expressed as a linear combinations of some measurable indicator functions, i.e.,

$$X = \sum_{1=1}^{n} a_i \mathbf{1}_{E_i} \tag{A.4}$$

for constants a_i and sets $E_i \in \mathcal{F}$. The expectation or mean of a simple random variable is denoted by $\mathbb{E}X$ and defined by

$$\mathbb{E}X = \sum_{1=1}^{n} a_i \mathbb{P}(E_i).$$

If X is a non-negative simple random variable, the expectation $\mathbb{E}X$ does not depend on the chosen representation (A.4).

If X is a non-negative (but not necessarily simple) random variable, i.e., $X \geq 0$, we define

$$\mathbb{E}X = \sup\{\mathbb{E}Y : Y \text{ is simple and } Y \leq X\}.$$

$\mathbb{E}X$ is always defined but may be equal to $+\infty$.

An arbitrary random variable can always be written as

$$X = X^+ - X^-,$$

where $X^+ = X\mathbf{1}_{[X \geq 0]}$ and $X^- = -X\mathbf{1}_{[X < 0]}$. Then $X^+, X^- \geq 0$ and $|X| = X^+ + X^-$.

We call X integrable if $\mathbb{E}|X| < \infty$, and we define

$$\mathbb{E}X = \mathbb{E}X^+ - \mathbb{E}X^-.$$

In particular, the expectation always exists if X is bounded, i.e., there exists some nonrandom constant $K > 0$ such that $|X| \leq K$.

We say that the expectation of X is the integral of X (which is a measurable function), with respect to $\mathbb{P}$ and we write

$$\mathbb{E}X = \int X \, d\mathbb{P} = \int X(\omega) \, d\mathbb{P}(\omega).$$

Note that $\mathbb{E}X$ is always defined when $X \geq 0$ almost surely. The latter concept meaning almost surely with respect to the probability measure $\mathbb{P}$. We say that a property holds *almost surely on a probability space* $(\Omega, \mathcal{F}, \mathbb{P})$ if it holds for every ω except for a set N such that $\mathbb{P}(N) = 0$. We abbreviate almost surely by a.s.

From the definition, one can deduce that the expectation is a linear and monotone operator, i.e., X, Y are integrable random variables and $a, b \in \mathbb{R}$, and then

1. $\mathbb{E}[aX + bY] = a\mathbb{E}X + b\mathbb{E}Y$;
2. if $X \geq Y$ a.s. then $\mathbb{E}X \geq \mathbb{E}Y$.

The main theorems of the Lebesgue integration theory follow.

Theorem A.10 (Monotone convergence) *Let X_n be a sequence of non-negative random variables such that $X_n \leq X_{n+1}$ a.s. for each n. Let $X = \lim X_n$ (we write also $X_n \nearrow X$), then*

$$\mathbb{E}X_n \nearrow \mathbb{E}X.$$

The next theorem is known as the dominated convergence theorem, also called Lebesgue's convergence theorem.

Theorem A.11 (Dominated convergence) *If (X_n), X, Y are random variables such that Y is integrable and non-negative, $|X_n| \leq Y$ for all n and $X_n \rightarrow X$ a.s., then X is integrable and $\mathbb{E}X_n \rightarrow \mathbb{E}X$.*

Theorem A.12 (Fatou lemma) *Let (X_n) be an arbitrary sequence of non-negative random variables. Then*

$$\mathbb{E}(\liminf X_n) \leq \liminf \mathbb{E}X_n.$$

If there exists a non-negative integrable random variable Y such that $X_n \leq Y$, then

$$\limsup \mathbb{E}X_n \leq \mathbb{E}(\limsup X_n).$$

The next two propositions have proven to be very useful in proofs of results in probability theory. The fact that $\mathbb{P}$ is a probability measure is essential.

Proposition A.13 *Let X be a real-valued random variable and $g : \mathbb{R} \rightarrow [0, \infty]$ an increasing function. Then*

$$\mathbb{P}(X \geq c)g(c) \leq \mathbb{E}g(X).$$

The inequality in Proposition A.13 is known as *Markov's inequality*. An example is obtained by taking $g(x) = x^+$ and replacing X by $|X|$. One gets

$$\mathbb{P}[|X| \geq c] \leq \frac{1}{c}\mathbb{E}|X|, \quad c > 0.$$

For the special case, when we have $g(x) = (x^+)^2$, the above inequality is known as the *Chebychev's inequality*. This name is particularly used when we apply the inequality with X replaced by $|X - \mathbb{E}X|$. Then, for $c > 0$, we obtain

$$\mathbb{P}\big(|X - \mathbb{E}X| \geq c\big) \leq \frac{1}{c^2}\,\mathrm{Var}\,X,$$

where the quantity $\mathrm{Var}\,X = \mathbb{E}(X - \mathbb{E}X)^2$ is known as the *variance* of X.

We now turn to a result that is known as *Jensen's inequality*.

Proposition A.14 *Let $g : G \to \mathbb{R}$ be a convex function (where G is a convex interval of $\mathbb{R}$). Suppose that X is a random variable such that $\mathbb{E}|X| < \infty$ and $\mathbb{E}|g(X)| < \infty$. Then*

$$\mathbb{E}\big(g(X)\big) \geq g\big(\mathbb{E}(X)\big).$$

Conditional Expectation   Let consider a random variable X that takes values in $\{x_1, x_2, \ldots, x_n\}$ and another one, Y, with values in $\{y_1, y_2, \ldots, y_m\}$. The conditional probability

$$\mathbb{P}\{X = x_i | Y = y_j\} = \frac{\mathbb{P}\{X = x_i, Y = y_j\}}{\mathbb{P}\{Y = y_j\}}$$

is well defined if $\mathbb{P}\{Y = y_j\} > 0$. Otherwise, we set it to zero. We write E_j for $\{Y = y_j\}$. The conditional expectation $\widehat{x}_j := \mathbb{E}[X|E_j]$ is then

$$\widehat{x}_j = \sum_i x_i \mathbb{P}\{X = x_i | E_j\}.$$

We define now a *new random variable* $\widehat{X}$ by

$$\widehat{X} := \sum_j \widehat{x}_j \mathbf{1}_{E_j}.$$

Since $\widehat{X} = \widehat{x}_j$ on each event $\{Y = y_j\}$, we call $\widehat{X}$ the *conditional expectation* of X given Y. Clearly, $\widehat{X}$ is $\sigma(Y)$-measurable. By simple calculations, it can be shown that

$$\mathbb{E}\widehat{X}\mathbf{1}_{E_j} = \mathbb{E}X\mathbf{1}_{E_j},$$

and then, since every event of $\sigma(Y)$ is a finite union of events E_j, we get

$$\mathbb{E}\widehat{X}\mathbf{1}_E = \mathbb{E}X\mathbf{1}_E, \quad \forall E \in \sigma(Y). \tag{A.5}$$

Moreover, $\widehat{X}$ is the only random variable that satisfies (A.5).

These properties are basic for defining a more general concept of conditional expectation of a random variable with respect to a σ-algebra.

Let $(\Omega, \mathcal{F}, \mathbb{P})$ be a probability space and $\mathcal{G}$ a sub-σ-algebra of $\mathcal{F}$. Assume that $X \in \mathcal{L}^1(\Omega, \mathcal{F}, \mathbb{P})$, i.e., X is an integrable random variable.

A random variable $\widehat{X}$ is called a version of the conditional expectation $\mathbb{E}[X|\mathcal{G}]$ if it is $\mathcal{G}$-measurable and

$$\mathbb{E}\widehat{X}\mathbf{1}_G = \mathbb{E}X\mathbf{1}_G, \quad \forall G \in \mathcal{G}.$$

If $\mathcal{G} = \sigma(Y)$, where Y is a random variable, instead of $\mathbb{E}[X|\sigma(Y)]$ we write $\mathbb{E}[X|Y]$.

Theorem A.15 *If $X \in \mathcal{L}^1(\Omega, \mathcal{F}, \mathbb{P})$, then a version of the conditional expectation* $\mathbb{E}[X|\mathcal{G}]$ *exists and, moreover, any two versions are a.s. equal.*

It is important to keep in mind that $\mathbb{E}[X|\mathcal{G}]$ is uniquely defined only up to almost sure equivalence.

Theorem A.16 (Properties of conditional expectation) *Suppose that* $X \in \mathcal{L}^1(\Omega, \mathcal{F}, \mathbb{P})$ *and* $\mathcal{G}, \mathcal{S}$ *are sub-σ-algebra of $\mathcal{F}$. Then:*

1. $\mathbb{E}[X|\mathcal{G}] = X$ *if X is $\mathcal{G}$-measurable.*
2. $\mathbb{E}[X|\mathcal{G}] = \mathbb{E}X$ *if X is independent of $\mathcal{G}$.*
3. $\mathbb{E}[XY|\mathcal{G}] = \mathbb{E}[X|\mathcal{G}]Y$ *if Y is bounded and $\mathcal{G}$-measurable.*
4. $\mathbb{E}[\mathbb{E}[X|\mathcal{S}]|\mathcal{G}] = \mathbb{E}[X|\mathcal{G}]$ *if $\mathcal{G} \subset \mathcal{S}$.*

If $\mathcal{G} = \{\Omega, \emptyset\}$ (the trivial σ-algebra) then the random variable with respect to $\mathcal{G}$ should be a constant a.s. Then

$$\mathbb{E}[X|\{\Omega, \emptyset\}] = \mathbb{E}X.$$

The conditional expectation is a monotone linear operator a.s.

A.4 Random Processes: General Theory

A random (or stochastic) process X is a family $\{X_t | t \in \mathcal{T}\}$ of random variables (in a general setting) that map the sample space Ω into some set $\mathbb{S}$. There are many possible choices for the index set $\mathcal{T}$ and the state space $\mathbb{S}$ and the process features depend strongly upon these choices. Usually, $\mathcal{T}$ is thought of as the time parameter. We may have *discrete time processes* when $\mathcal{T} = \{0, 1, 2, \ldots\}$ or *continuous time processes* when $\mathcal{T} = [0, \infty)$. Other choices for $\mathcal{T}$ include $\mathbb{R}^n$ and $\mathbb{Z}^n$. The state space $\mathbb{S}$ might be a countable set like $\mathbb{Z}$ or an uncountable set like $\mathbb{R}$. The analysis of random processes varies depending in a great measure on whether $\mathbb{S}$ and $\mathcal{T}$ are countable or uncountable, in the same way that discrete random variables are distinguishable from continuous variables.

There exist two perspectives for describing the evolution of a random process X:

1. Each X_t is a function that maps Ω into $\mathbb{S}$. For any fixed elementary event $\omega \in \Omega$, there is a corresponding collection $\{X_t(\omega) | t \in \mathcal{T}\}$ of elements of $\mathbb{S}$. This is called the *realisation* or *sample path* or *trajectory* of X at ω. Sample paths are subject to different studies.
2. In general, the random variables X_t are not independent. If $\mathbb{S} \subset \mathbb{R}$ and $\mathbf{t} = (t_1, t_2, \ldots, t_n)$ is a vector of members of $\mathcal{T}$, then the vector $(X_{t_1}, X_{t_2}, \ldots, X_{t_n})$ has the joint distribution $F_{\mathbf{t}} : \mathbb{R}^n \to [0, 1]$ given by

$$F_{\mathbf{t}}(x_1, x_2, \ldots, x_n) = \mathbb{P}(X_{t_1} \leq x_1, \ldots, X_{t_n} \leq x_n).$$

The collection $\{F_{\mathbf{t}} | \mathbf{t} \text{ is vector of finite length of elements of } \mathcal{T}\}$ is called the collection of *finite-dimensional distributions* of X and it provides all the information about X derived from the distributions of its component variables X_t. Then the distributional properties of X can be studied using its finite-dimensional distributions.

A fundamental theorem of Kolmogorov states that given any consistent family of finite-dimensional distributions there exists a random process on an appropriate probability space such that its finite-dimensional distributions coincide with the given ones.

The process X is *continuous* (*right, left continuous*) if its trajectories have these properties a.s. If the trajectories are right continuous with left limits (rcll), we say that the process has the *càdlàg* property. Càdlàg is the abbreviation for 'continue à la droite avec limits à la gauche', the French translation of 'right continuous with left limits'.

Measurability   Let us consider the case of a continuous time real-valued random process, when $\mathcal{T} = [0, \infty)$. The process X is called *measurable* if as a function $X(t, \omega)$ on the product space

$$X : [0, \infty) \times \Omega \to \mathbb{R}$$

is measurable with respect to the product σ-algebra $\mathcal{B}(\mathbb{R}) \otimes \mathcal{F}$. If X is measurable, then for every $\omega \in \Omega$ the function $t \to X(t, \omega)$ is measurable and we can define new random variables like

$$\int_0^t g\big(X(t, \omega)\big)\, ds,$$

where g is a bounded measurable function.

Two processes X and X' are:

- *modifications* of each other if

$$\mathbb{P}\big(X_t = X_t'\big) = 1, \quad \forall t \geq 0;$$

in this case the processes have the same infinite-dimensional distributions;
- *indistinguishable* if the sample paths

$$t \to X(t, \omega) \quad \text{and} \quad t \to X'(t, \omega)$$

are identical for all $\omega \in G$ with $P(G) = 1$.

Natural Filtration   Let $\{\mathcal{F}_t\}$ be an increasing family of sub-σ-algebras of $\mathcal{F}$, i.e., $\mathcal{F}_s \subset \mathcal{F}_t$ if $s \leq t$. A *filtered probability space* $(\Omega, \mathcal{F}, \mathcal{F}_t, \mathbb{P})$ is a probability space $(\Omega, \mathcal{F}, \mathbb{P})$ together with a filtration $\{\mathcal{F}_t\}$. We say that X is an *adapted process* if for all $t \geq 0$ the random variable X_t is $\mathcal{F}_t$-measurable. In this case, $\{\mathcal{F}_t\}$ is called an admissible filtration for X.

Let $\mathcal{F}_t^{0\,X}$ be the smallest σ-algebra in $\mathcal{F}$ with respect to which all the random variables $\{X_s | 0 \leq s \leq t\}$ are measurable. If $t_1 \leq t_2$ then $\mathcal{F}_{t_1}^{0\,X} \subset \mathcal{F}_{t_2}^{0\,X}$ (using the definition), i.e., $\{\mathcal{F}_t^{0\,X}\}_{t\geq 0}$ is an increasing family of sub-σ-algebras of $\mathcal{F}$. It is called the *natural filtration* of X. For each $t \geq 0$, the tuple $(\Omega, \mathcal{F}_t^{0\,X}, \mathbb{P})$ is a probability space. Usually, in practice, we work with $(\mathcal{F}, \mathbb{P})$-*completion* of $\mathcal{F}_t^{0\,X}$, which is the σ-algebra of subsets of Ω expressible in the form $A = A_1 \cup A_2$, where $A_1 \in \mathcal{F}_t^{0\,X}$ and $A_2 \subset A_3$ with $A_3 \in \mathcal{F}$ and $\mathbb{P}(A_3) = 0$.

$\{\mathcal{F}_t\}$ is *right continuous* if $\mathcal{F}_t = \bigcap_{s>t} \mathcal{F}_s$.

Stopping Times   A positive random variable $T : \Omega \to [0, \infty)$ is called a *stopping time* with respect to $\{\mathcal{F}_t\}$ provided $(T \leq t) \in \mathcal{F}_t$ for all $t \geq 0$.

A σ-algebra $\mathcal{F}_t$ should be thought of as a 'measure' of the information available at time t, i.e., all events of $\mathcal{F}_t$ are observable at time t, in the sense that we know whether they are realised or not at time t.

A stopping time T can be associated with the σ-algebra of the events prior to T, denoted by $\mathcal{F}_T$, i.e.,

$$A \in \mathcal{F}^T \iff A \cap \{T \leq t\} \in \mathcal{F}_t, \quad \forall t \geq 0.$$

Some very useful properties of stopping times are provided below:

- The minimum, maximum and sum of two stopping times is still a stopping time.
- If $S \geq T$ and S is $\mathcal{F}_T$-measurable, where T is a stopping time, then S is also a stopping time.
- If T is a stopping time, then T is $\mathcal{F}_T$-measurable.
- If S, T are two stopping times and $A \in \mathcal{F}_S$, then $A \cap \{S \leq T\} \in \mathcal{F}_T$.
- If S, T are two stopping times such that $S \leq T$ then $\mathcal{F}_S \subset \mathcal{F}_T$.
- If S, T are two stopping times then $\{S < T\}$, $\{S = T\}$, $\{S > T\}$ belong to $\mathcal{F}_T \cap \mathcal{F}_S$.
- If ξ is $\mathcal{F}_S$-measurable and η is $\mathcal{F}_T$-measurable, then $\{\xi = \eta\} \cap \{S = T\} \in \mathcal{F}_T \cap \mathcal{F}_S$.
- The supremum of a sequence of stopping times is still a stopping time.
- The limit of an increasing sequence of stopping times is a stopping time.

Suppose that $\{X_t\}$ is a stochastic process with values in $\mathbb{R}^n$ which is right continuous and adapted with respect to a complete right continuous filtration $\{\mathcal{F}_t\}$. Let $A \subset \mathbb{R}^n$ be a Borel-measurable set. We define the functions

$$D_A(\omega) = \inf\{t \geq 0 : X_t(\omega) \in A\},$$

$$T_A(\omega) = \inf\{t > 0 : X_t(\omega) \in A\},$$

where in all cases the infimum of the empty set is understood to be $+\infty$. We call D_A (respectively, T_A) the *first entry* (respectively, *hitting*) *time* of A. Then D_A and T_A are the most useful and standard examples of stopping times.

A.5 Stochastic Models

A.5.1 Martingales

Martingales in discrete and continuous time play an important role in many areas of probability including convergence of stochastic processes (or invariance principles) based on Skorokhod's embedding, stochastic integration, stochastic differential equations, central limit theorems, stochastic calculus of point processes and diffusion processes.

Informally a martingale is simply a real-valued stochastic process M_t defined on some probability space $(\Omega, \mathcal{F}, \mathbb{P})$ that is conditionally constant, i.e., whose predicted value at any future time $s > t$ is the same as its present value at the time t of prediction.

The set $\mathcal{T}$ of possible indices is usually taken to be the non-negative integers $\mathbb{Z}_+$ or the non-negative reals $\mathbb{R}_+$, although sometimes $\mathbb{Z}$ or $\mathbb{R}$ or other ordered sets arise. Formally we represent what is known at time t in the form of an increasing family (or filtration) $\{\mathcal{F}_t\} \subset \mathcal{F}$ of σ-algebras, possibly those generated by a process $\{X_s : s \leq t\}$ or even by the martingale itself, $\mathcal{F}_t = \sigma\{M_s : s \leq t\}$ and require that $\mathbb{E}[M_t] < \infty$ for each t (so the conditional expectation below is well defined) and that

$$M_t = \mathbb{E}[M_s | \mathcal{F}_t], \quad t < s \text{ a.s.}$$

In particular, $\{M_t\}$ is adapted to $\{\mathcal{F}_t\}$. $\{M_t\}$ is supermartingale (respectively, submartingale) if

$$M_t \geq (\text{respectively, } \leq) \mathbb{E}[M_s | \mathcal{F}_t], \quad t < s \text{ a.s.}$$

Using the properties of conditional expectations, one can show that a martingale has a constant expectation, i.e.,

$$\mathbb{E}M = \mathbb{E}M_0 \quad \text{for all } t \geq 0,$$

whilst a super(sub)martingale has a monotone decreasing (increasing) expectation. Note that $-M_t = (-M_t, t \geq 0)$ is a supermartingale if M is a submartingale and M is a martingale if and only if it is simultaneously a super- and submartingale.

In a certain sense, martingales are the constant functions of probability theory; submartingales are the increasing functions and supermartingales are the decreasing functions. With this interpretation, if X_t stands for the fortune of a gambler at time t, then a martingale (submartingale, supermartingale) corresponds to the notion of a fair (respectively, favourable, unfavourable) game.

Estimates on Martingales   If T is a stopping time and if M_t is a martingale, then $M_{t \wedge T}$ is a martingale too.

Theorem A.17 *Let $\{M_t\}$ be a right continuous submartingale with respect to the filtration $\{\mathcal{F}_t\}$. Then the following results hold:*

(i) Doob's Inequality:

$$\mathbb{P}\left\{\omega \in \Omega \;\middle|\; \sup_{0 \le s \le t} |M_s| \ge \lambda\right\} \le \frac{\mathbb{E}|M_t|}{\lambda}, \quad \lambda > 0.$$

(ii) Doob's Maximal Inequality:

$$\mathbb{E} \sup_{0 \le s \le t} |M_s| \le \left(\frac{\alpha}{\alpha - 1}\right)^{\alpha} \mathbb{E}|M_t|^{\alpha},$$

provided that $\alpha > 1$ and $M_s \in L^{\alpha}(\Omega, \mathcal{F}, \mathbb{P})$.

Theorem A.18 (Optimal sampling (stopping) theorem) *Let $\{M_t\}$ be a right contin- uous submartingale with respect to the filtration $\{\mathcal{F}_t\}$ with the last element M_∞. If S, T are two stopping times such that $S \le T$, then*

$$\mathbb{E}[M_T \,|\, \mathcal{F}_S] \ge M_S \quad a.s.$$

In particular, $\mathbb{E}[M_T] \ge \mathbb{E}[M_0]$ and for a martingale with last element $\mathbb{E}[M_T] = \mathbb{E}[M_0]$.

Martingale Path Regularity

Theorem A.19 *Suppose that M is a submartingale with respect to the filtration $\{\mathcal{F}_t\}$ where is right continuous and complete. Then M has a right continuous modi- fication if and only if $t \mapsto \mathbb{E}M_t$ is right continuous. This modification can be chosen to have the càdlàg property and to be adapted to $\{\mathcal{F}_t\}$.*

A.5.2 Markov Processes

Markov processes have the following property: Given that its current state is known, the probability of any future event of the process is not altered by additional knowl- edge concerning its past behaviour. Formally, a stochastic dynamical system sat- isfies the Markov property (formulated by A.A. Markov in 1906) if the probable (future) state of the system at any time $t > s$ is independent of the (past) behaviour of the system at times $t < s$, given the present state at time s. This is the stochas- tic analog of an important property shared with solutions of initial-value problems involving ODEs and so stochastic processes satisfying this property arise naturally.

   Let consider the case in which the time is discrete and we have a stochastic process $\{x_n\}_{n \in \mathbb{N}}$. At an instant of time n, the process could be in any of a countable number of states $\{0, 1, 2, \ldots, p, \ldots\}$. If $x_n = i$ is the system state at instant n, at the next time instant $n + 1$, the system would be in state j with a probability p_{ij} such that

$$p_{ij} = \mathbb{P}\{x_{n+1} = j \,|\, x_n = i, x_{n-1} = k, \ldots, x_0 = l\} = \mathbb{P}\{x_{n+1} = j \,|\, x_n = i\}.$$

This relation illustrates the memoryless property of the chain, since p_{ij} is independent of the past (it does not depend on how the process arrived in state i in the first place).

One can consider a stochastic process taking values in a measurable space $(\mathbb{S}, \mathcal{B})$, called the *state space*. If $\mathbb{S}$ is a Hausdorff (or separated) topological space we denote by $\mathcal{B}(X)$ or $\mathcal{B}$ its Borel σ-algebra.

Formally, a stochastic process $X = \{X_t | t \in \mathbb{R}_+\}$ with the state $(\mathbb{S}, \mathcal{B})$, defined on a filtered probability space $(\Omega, \mathcal{F}, \mathcal{F}_t, \mathbb{P})$ is a Markov process if, for any times t, s with $t \geq s$ and any bounded measurable function $f : \mathbb{S} \to \mathbb{R}$, the following equality holds:

$$\mathbb{E}\big[f(x_t)\big|\mathcal{F}_s\big] = \mathbb{E}\big[f(x_t)\big|x_s\big],$$

where $\mathbb{E}$ is the expectation with respect to $\mathbb{P}$. This says that the only information relevant to evaluating the behaviour of the process beyond time s is the value of the current state, x_s. It implies, in particular, that X is adapted to $\mathcal{F}_t$.

Poisson Processes   A Poisson distribution with parameter $\mu > 0$ is defined as

$$p_k = \frac{e^{-\mu} \mu^k}{k!}$$

and describes the probability that k events happened over a time period embedded in μ. If a random variable has a Poisson distribution then

$$\mathbb{E}[X] = \mu, \quad \mathrm{Var}[X] = \mu.$$

A Poisson process is a special type of Markov process that comprises concepts of Poisson distribution together with independence. A Poisson process of intensity $\lambda > 0$ that describes the expected number of events per time unit is a stochastic process $\{X(t) | t \geq 0\}$ with values in the set of integers $\mathbb{Z}$ such that the following axioms hold:

1. For any sequence of times $t_0 = 0 < t_1 < t_2 < \cdots < t_n$, the number of events that take place in disjoint intervals (process increments)

$$X(t_1) - X(t_0), \quad X(t_2) - X(t_1), \quad \ldots, \quad X(t_n) - X(t_{n-1})$$

   are independent random variables. This property is known as the *independent increments property* of the Poisson process.

2. If $s \geq 0$ and $t > 0$ then the random variable $X(s + t) - X(s)$ follows the Poisson distribution

$$\mathbb{P}\big(X(s + t) - X(s) = k\big) = \frac{(\lambda t)^k \exp(-\lambda t)}{k!}.$$

3. $X(0) = 0$, i.e., at time zero the number of events that have already happened is zero.

For a Poisson process with a rate $\lambda > 0$, we have

$$\mathbb{E}\big[X(t)\big] = \lambda t, \quad \mathrm{Var}\big[X(t)\big] = \lambda t.$$

For a small interval of time h, we have

$$\mathbb{P}\big(X(t+h) - X(t) = 1\big) = \frac{(\lambda h)\exp(-\lambda h)}{1!} = \lambda h + o(h).$$

Therefore, the rate λ is the proportionality constant in the probability that one event will occur during an arbitrary small interval h. We denote by w_i the occurrence time of the ith event. The waiting time between consecutive events is called the *sojourn time*,

$$S_i = w_{i+1} = w_i,$$

and it represents the holding time of the Poisson process in the state i. The sojourn times $S_0, S_1, \ldots, S_{n-1}$ are independent random variables, each having the exponential probability density function

$$f_{S_k}(s) = \lambda \exp(-\lambda s).$$

Birth-and-Death Processes   The birth-and-death process is a special case of Markov process, in which the states represent the current size of a population, and the transitions are limited to birth and death. The process transitions are of types (a) $i \to i + 1$ when a birth occurs and (b) $i \to i - 1$ when a death occurs. The birth-and-death events are assumed to be independent of each other. The birth-and-death process is characterised by the birth rate $\{\lambda_i\}_{i=0,1,\ldots,\infty}$ and the death rate $\{\mu_i\}_{i=0,1,\ldots,\infty}$, which depend on each state i.

The description of a birth-and-death process is as follows. After the process enters in state i, it stays there for some random length of time, exponentially distributed with parameter $(\lambda_i + \mu_i)$. When leaving state i, the process enters either $i + 1$ with probability $\frac{\lambda_i}{\lambda_i + \mu_i}$ or $i - 1$ with probability $\frac{\mu_i}{\lambda_i + \mu_i}$. Then in the state $i + 1$ the story repeats.

Let us consider two random variables $B(i)$ and $D(i)$ exponentially distributed with parameters λ_i and μ_i, respectively. $B(i)$ is thought of as the time until a birth occurs and $D(i)$ describes the time until a death occurs when the population size is i. If $B(i)$ and $D(i)$ are independent random variables, then their minimum (the sojourn time in the state i) is again a random variable exponentially distributed with the parameter $(\lambda_i + \mu_i)$. A transition from i to $i + 1$ takes place provided that $B(i) < D(i)$ and

$$\mathbb{P}\big[B(i) < D(i)\big] = \frac{\lambda_i}{\lambda_i + \mu_i}.$$

A stochastic process $\{X(t)|t \geq 0\}$ is called a birth-and-death process, if for a very short time interval $h \searrow 0$, the following axioms hold:

1. $\mathbb{P}(X(t+h) - X(t) = 1|X(t) = i) = \lambda_i h + o(h)$,
2. $\mathbb{P}(X(t+h) - X(t) = -1|X(t) = i) = \mu_i h + o(h)$,
3. $\mathbb{P}(X(t+h) - X(t) > 1|X(t) = i) = o(h)$,
4. $\mu_0 = 0, \lambda_0 > 0, \mu_i, \lambda_i > 0$ if $i \geq 1$.

The above postulates imply that

$$\mathbb{P}\big(X(t+h) - X(t) = 0 \big| X(t) = i\big) = 1 - (\mu_i + \lambda_i)h + o(h).$$

The description of a birth-and-death process shows that we can identify two Poisson processes embedded in the whole process: one called *pure birth process* with birth rate $\{\lambda_i\}_{i=0,1,\dots,\infty}$ and one called *pure death process* with death rate $\{\mu_i\}_{i=0,1,\dots,\infty}$.

Wiener Processes   The Wiener process (or Brownian motion) is by far the most interesting and fundamental stochastic process. It was studied by A. Einstein (1905) in the context of a kinematic theory for the irregular movement of pollen immersed in water, first observed by the botanist R. Brown in 1824 and later by Bachelier (1900) in the context of financial economics. Its mathematical theory was initiated by N. Wiener (1923). P. Lévy (1948) became famous by studying its sample paths. His work was an inspiration for practically all subsequent research on stochastic processes until today. Appropriately, the process is also known as the Wiener–Lévy process and finds applications in engineering (communications, signal processing, control), economics and finance, mathematical biology, management science, etc. Without going into the construction of the Wiener process, we can point out the properties that characterise it. The two key properties of the Wiener process $\{W_t\}$ are:

1. *Gaussian increments.* For each $s < t$ the random variable $W_t - W_s$ has a normal (Gaussian) distribution with mean zero and variance $\sigma^2(t - s)$. Therefore, it is a particular *Lévy process*.
2. *Independent increments.* The increments $W_t - W_s$ corresponding to disjoint intervals are independent. This involves also the memoryless feature of the Wiener process.

The first property implies the distribution of the increment $W_t - W_s$ depends only on $t - s$. This is called the *stationarity (invariance) of the increments*.

It is often convenient to specify that the Wiener process has a particular value at $t = 0$, for instance $W(0) = 0$. Sometimes we may specify another value. It is also convenient to think of the Wiener process as defined for all real t. This can be arranged by imagining another Wiener process going backwards in time and joining the two at time zero. The trajectories (paths) of the Wiener process are continuous.

Many times in this book we use the term *standard Wiener process*, or *Brownian motion*. This will be formally defined as follows.

The stochastic process $W = \{W_t : t \geq 0\}$ is called a (standard) Brownian motion or Wiener process if

(i) $W(0) = 0$ a.s.,
(ii) $W_t - W_s$ is independent of $\{W_u : u \leq \cdot s\}$ for all $s \leq t$,
(iii) $W_t - W_s$ has an $N(0, t - s)$-distribution for all $s \leq t$,
(iv) almost all sample paths of W are continuous.

It can be proved that such a process really exists. The sample paths are continuous but 'very rough'. This is illustrated by the following classical result.

Theorem A.20 *Let W_t be a Brownian motion. For all ω outside a set of probability zero, the sample path $t \rightarrow W_t(\omega)$ is nowhere differentiable.*

Computation of Joint Probabilities   From the definition of the Brownian motion, we know

$$\mathbb{P}\big(a \leq W(t) \leq b\big) = \frac{1}{\sqrt{2\pi t}} \int_a^b e^{-\frac{x^2}{2t}} \, dx$$

for all $t > 0$ and $a \leq b$ since $W(t) \sim N(0, t)$.

Problem Suppose we choose times $0 < t_1 < \cdots < t_n$ and real numbers $a_i \leq b_i$ for $i = 1, \ldots, n$. What is the joint probability

$$\mathbb{P}\big(a_1 \leq W(t_1) \leq b_1, \ldots, a_n \leq W(t_n) \leq b_n\big)?$$

Step 1.

$$\mathbb{P}\big(a_1 \leq W(t_1) \leq b_1\big) = \frac{1}{\sqrt{2\pi t_1}} \int_{a_1}^{b_1} \exp\left[-\frac{x_1^2}{2t_1}\right] dx_1.$$

Given that $W(t_1) = x_1$, $a_1 \leq x_1 \leq b_1$, then presumably the process is $N(x_1, t_2 - t_1)$ on the interval $[t_1, t_2]$. Thus the probability that $a_2 \leq W(t_2) \leq b_2$, given that $W(t_1) = x_1$, should be

$$\frac{1}{\sqrt{2\pi(t_2 - t_1)}} \int_{a_2}^{b_2} \exp\left[-\frac{(x_2 - x_1)^2}{2(t_2 - t_1)}\right] dx_2.$$

Then

$$\mathbb{P}\big(a_1 \leq W(t_1) \leq b_1, a_2 \leq W(t_2) \leq b_2\big)$$
$$= \int_{a_1}^{b_1} \int_{a_2}^{b_2} g(x_1, t_1 | 0) g(x_2, t_2 - t_1 | x_1) \, dx_2 \, dx_1$$

for

$$g(x, t | y) := \frac{1}{\sqrt{2\pi t}} \exp\left[-\frac{(x - y)^2}{2t}\right].$$

Step 2.

$$\mathbb{P}\big(a_1 \leq W(t_1) \leq b_1, \ldots, a_n \leq W(t_n) \leq b_n\big)$$
$$= \int_{a_1}^{b_1} \cdots \int_{a_n}^{b_n} g(x_1, t_1 | 0) \ldots g(x_n, t_n - t_{n-1} | x_{n-1}) \, dx_n \ldots dx_1.$$

Theorem A.21 *Let $W(\cdot)$ be a one-dimensional Wiener process. Then for all positive integers n, all choices of times $0 = t_0 < t_1 < \cdots < t_n$ and each function $f : \mathbb{R}^n \rightarrow \mathbb{R}$, we have*

$$E\big[f\big(W(t_1),\ldots,W(t_n)\big)\big]$$
$$= \int_{-\infty}^{+\infty} \cdots \int_{-\infty}^{+\infty} f(x_1,\ldots,x_n)g(x_1,t_1\,|\,0)\ldots g(x_n, t_n - t_{n-1}\,|\,x_{n-1})\,dx_n \ldots dx_1$$

for

$$g(x,t\,|\,y) := \frac{1}{\sqrt{2\pi t}}\exp\left[-\frac{(x-y)^2}{2t}\right].$$

Quadratic Variation   Despite their continuity, the sample paths $t \mapsto W_t(\omega)$ are not differentiable anywhere on $[0,\infty)$. Fix $t > 0$ and take a sequence of partitions $0 = t_0^{(n)} < t_1^{(n)} < \cdots < t_k^{(n)} < \cdots < t_{2^n}^{(n)} = t$ of the interval $[0,t]$, such that $t_k^{(n)} = kt2^{-n}$. We consider the variation of order p of the sample path $t \mapsto W_t(\omega)$ along the nth partition as follows:

$$V_p^{(n)}(\omega) := \sum_{k=1}^{2^n} \big|W_{t_k^{(n)}}(\omega) - W_{t_{k-1}^{(n)}}(\omega)\big|^p, \quad p > 0.$$

When $p = 1$, $V_p^{(n)}$ is just the length of the polygonal approximation to the Wiener process. When $p = 2$, $V_p^{(n)}$ is the quadratic variation of the path along the approximation.

Theorem A.22 *With probability one, we have the following result:*

$$\lim_{n\to\infty} V_p^{(n)} = \begin{cases} \infty, & \text{if } p \in (0,2) \\ t, & \text{if } p = 2 \\ 0, & \text{if } p > 2. \end{cases}$$

Stochastic Integral   From the practitioner's point of view, Itô calculus is a tool for manipulating those stochastic processes which are most closely related to Brownian motion.

The central result of this theory is the famous Itô formula. The understanding of the Itô formula involves the manipulation of the *Itô stochastic integral*. The Itô integral is a mathematical object which is only roughly analogous to the traditional integral of Newton and Leibniz. The real driving force behind the definition and the effectiveness of the Itô integral is that it carries the notion of the martingale transform from discrete time into continuous time. This type of integral provides the modeller with new tools for specifying stochastic processes in terms of 'differentials'.

Suppose that W is a standard Wiener process adapted to a given filtration $\{\mathcal{F}_t\}$; for a suitable adapted process X, we aim to define the following stochastic integral:

$$I_t(X) = \int_0^t X_s \, dW_s$$

and to study the path properties of the stochastic process (I_t). Since the first variation on any interval $[0,t]$ of the path $s \mapsto W_s(\omega)$ is infinite, defining the (I_t) as a

Lebesgue–Stieltjes integral is not possible. Therefore, a new approach is necessary such that the fact that the path has finite and positive quadratic variation could be used. In the following, we will discuss the concept of a stochastic integral leaving aside some technicalities that are required to make our definitions rigorous.

We say that the process X is *simple or elementary* if it is piecewise constant, i.e., there exists a sequence of stopping times $0 = t_0 < t_1 < \cdots < t_r < t_{r+1} = T$ such that $X_s(\omega) = e_j(\omega)$; $t_j < s < t_{j+1}$ where e_j is bounded, $\mathcal{F}_{t_j}$-measurable random variable. For such a process, the stochastic integral is defined as follows:

$$I_t(X) := \sum_{j=0}^{r} e_j(W_{t \wedge t_{j+1}} - W_{t \wedge t_j}).$$

For a more general process, X_t, we have

$$\int_0^t X_s \, dW_s := \lim_{n \to \infty} \int_0^t X_s^{(n)} \, dW_s$$

where $X_t^{(n)}$ is a sequence of simple processes that converges in a suitable way to X_t.

We define the space $L^2[0, T]$ to be the space of mean square-integrable processes X_t, i.e.,

$$\mathbb{E}\left[\int_0^T X_t(\omega)^2 \, dt \right] < \infty.$$

Theorem A.23 (Itô isometry) *For any $X_t \in L^2[0, T]$, we have*

$$\mathbb{E}\left[\left(\int_0^T X_t(\omega) \, dW_t \right)^2 \right] = \mathbb{E}\left[\int_0^T X_t(\omega)^2 \, dt \right].$$

Theorem A.24 (Martingale property of stochastic integrals) *The stochastic integral $I_t(X) = \int_0^t X_s \, dW_s$ is a martingale when X_t is in $L^2[0, T]$.*

The Chain Rule in Stochastic Calculus   Assume that $n = 1$ and $X(\cdot)$ solves the following stochastic differential equation:

$$dX = b(X) \, dt + dW;$$

i.e.,

$$X_t = X_0 + \int_0^t b(X_s) \, ds + \int_0^t dW_s.$$

Suppose that $u : \mathbb{R} \to \mathbb{R}$ is a given smooth function. We ask: What stochastic differential equation does

$$Y(t) := u\big(X(t)\big), \quad t > 0$$

solve? One might use the usual chain rule to write

$$dY = u' \, dX = u' b \, dt + u' \, dW$$

where $' = \frac{d}{dx}$. This is *wrong*! In fact, in some sense we have

$$dW \approx (dt)^{1/2}.$$

Consequently, if we compute dY and keep all terms of order dt or $(dt)^{1/2}$, we obtain

$$dY = u' \, dX + \frac{1}{2} u''(dX)^2 + \cdots$$

$$= u'(b \, dt + dW) + \frac{1}{2} u''(b \, dt + dW)^2 + \cdots$$

$$= \left(u'b + \frac{1}{2} u'' \right) dt + u' \, dW + \{\text{terms of order} \geq (dt)^{3/2}\}.$$

Hence

$$dY = \left(u'b + \frac{1}{2} u'' \right) dt + u' \, dW$$

with an extra term $\frac{1}{2} u'' \, dt$ not present in ordinary calculus.

Theorem A.25 (Chain rule) *Let $u : \mathbb{R} \to \mathbb{R}$ be a twice-differentiable function and W be the standard one-dimensional Brownian motion. Then*

$$u(W_t) = u(W_0) + \int_0^t f'(W_s) \, dW_s + \frac{1}{2} \int_0^t f''(W_s) \, ds, \quad 0 \leq t < \infty.$$

References

1. Abate, A., D'Innocenzo, A., Benedetto, M.D., Sastry, S.: Markov set-chains as abstractions of stochastic hybrid systems. In: Egerstedt, M., Mishra, B. (eds.) Hybrid Systems: Computation and Control. Lecture Notes in Computer Science, vol. 4981, pp. 1–15. Springer, Berlin (2008)
2. Abate, A., Prandini, M., Lygeros, J., Sastry, S.: Probabilistic reachability and safety for controlled discrete time stochastic hybrid systems. Automatica **44**(11), 2724–2734 (2008)
3. Abate, A., Prandini, M., Lygeros, J., Sastry, S.: Approximation of general stochastic hybrid systems by switching diffusions with random hybrid jumps. In: Egerstedt, M., Mishra, B. (eds.) Hybrid Systems: Computation and Control. Lecture Notes in Computer Science, vol. 4981, pp. 598–601. Springer, Berlin (2008)
4. Abate, A., Katoen, J.-P., Prandini, M., Lygeros, J.: Approximate model checking of stochastic hybrid systems. Eur. J. Control **6**, 624–641 (2010)
5. Albeverio, S., Ma, M.: A general correspondence between Dirichlet forms and right processes. Bull. Am. Math. Soc. **26**(2), 245–252 (1992)
6. Alur, R., Henzinger, T.A., Ho, P.-H.: Automated symbolic verification of embedded systems. IEEE Trans. Softw. Eng. **22**(3), 181–201 (1996)
7. Arnold, L.: Stochastic Differential Equations: Theory and Applications. Wiley-Interscience, New York (1974)
8. Arrowsmith, D.K., Place, C.M.: Dynamical Systems: Differential Equations, Maps and Chaotic Behavior. Chapman & Hall, London (1992)
9. Aziz, A., Sanwal, K., Singhal, V., Brayton, R.: Model checking continuous time Markov chains. ACM Trans. Comput. Log. **1**(1), 162–170 (2000)
10. Bachelier, L.: Théorie de la Speculation. Gauthier-Villars, Paris (1900)
11. Barles, C., Chasseigne, E., Imbert, C.: On the Dirichlet problem for second-order elliptic integro-differential equations. Preprint (2007)
12. Bassan, B., Ceci, C.: Regularity of the value function and viscosity solutions in optimal stopping problems for general Markov processes. Stoch. Stoch. Rep. **74**(3–4), 633–649 (2002)
13. Baier, C., Katoen, J.-P., Hermanns, H.: Approximate symbolic model checking of continuous time Markov chains. In: Baeten, J.C.M., Mauw, S. (eds.) Concurrency Theory. Lecture Notes in Computer Science, vol. 1664, pp. 146–162. Springer, Berlin (1999)
14. Baier, C., Haverkort, B.R., Hermanns, H., Katoen, J.-P.: Reachability in continuous-time Markov reward decision processes. In: Logic and Automata: History and Perspectives, pp. 53–71 (2007)
15. Baxter, J.R., Chacon, R.V.: Compactness of stopping times. Z. Wahrscheinlichkeitstheor. Verw. Geb. **40**(3), 169–181 (1977)
16. Bect, J.: A unifying formulation of the Fokker–Planck–Kolmogorov equation for general stochastic hybrid systems. Nonlinear Anal. Hybrid Syst. **4**(2), 357–370 (2010)

17. Bect, J., Baili, H., Fleury, G.: Generalized Fokker–Planck equation for piecewise-diffusion processes with boundary hitting resets. In: Proceedings of the 17th International Symposium on the Mathematical Theory of Networks and Systems (2006)

18. Bemporad, A., Di Cairano, S.: Modelling and optimal control of hybrid systems with event uncertainty. In: Morari, M. Thiele, L. (eds.) Hybrid Systems: Computation and Control. Lecture Notes in Computer Science, vol. 3414, pp. 151–167. Springer, Berlin (2005)

19. Bensoussan, A., Menaldi, J.L.: Stochastic hybrid control. J. Math. Anal. Appl. **249**, 261–288 (2000)

20. Benjamini, I., Pemantle, R., Peres, Y.: Martin capacity for Markov chains. Ann. Probab. **23**, 1332–1346 (1995)

21. Berger, J.O.: Statistical Decision Theory and Bayesian Analysis. Springer, Berlin (1985)

22. Bertsekas, D.P., Shreve, S.E.: Stochastic Optimal Control: The Discrete-Time Case. Athena Scientific, Nashua (1996)

23. Beurling, A., Deny, J.: Dirichlet spaces. Proc. Natl. Acad. Sci. USA **45**, 208–215 (1959)

24. Biere, A., Cimatti, A., Clarke, E., Strichman, O., Zhu, Y.: Bounded model checking. Adv. Comput. **58**, 118–149 (2003)

25. Bismut, J.-M., Skalli, B.: Temps d'arrêt optimal, théorie générale des processus et processus de Markov. Probab. Theory Relat. Fields **36**(4), 301–313 (1977)

26. Blom, H.A.P.: Stochastic hybrid processes with hybrid jumps. In: Analysis and Design of Hybrid System, pp. 319–324. IFAC Press, New York (2003)

27. Blom, H.A.P.: Bayesian estimation for decision—directed stochastic control. PhD Thesis, Delft University of Technology (1990)

28. Blom, H.A.P., Lygeros, J. (eds.): Stochastic Hybrid Systems: Theory and Safety Critical Applications. Lecture Notes in Control and Information Sciences, vol. 337. Springer, Berlin (2006)

29. Blom, H.A.P., Bakker, G.J., Krystul, J.: Probabilistic reachability analysis for large scale stochastic hybrid systems. In: Proc. of the 46th IEEE Conference on Decision and Control, pp. 3182–3189 (2007)

30. Blom, H.A.P., Bakker, G.J., Krystul, J., Klompstra, M.B., Obbink, B.K.: Free flight collision risk estimation by sequential Monte Carlo simulation. In: Cassandras, C.G., Lygeros, J. (eds.) Stochastic Hybrid Systems; Recent Developments and Research Trends. Taylor & Francis/CRC Press, London/Boca Raton (2007) (Chap. 10)

31. Blom, H.A.P., Bakker, G.J.: Conflict probability and incrossing probability in air traffic management. In: Proc. of the 41st IEEE Conference on Decision and Control 3, 2421–2426 (2002)

32. Blom, H.A.P., Stroeve, S.H., Everdij, M.H.C., Van der Park, M.N.J.: Collision risk modelling of air traffic. In: Proc. of European Control Conference (2003)

33. Blumenthal, R.M., Getoor, R.K.: Markov Processes and Potential Theory. Academic Press, New York (1968)

34. Borkar, V.S., Ghosh, M.K., Sahay, P.: Optimal control of a stochastic hybrid system with discounted cost. J. Optim. Theory Appl. **101**(3), 557–580 (1999)

35. Bouleau, N., Lepingle, D.: Numerical Methods for Stochastic Processes. Wiley Series in Probability and Mathematical Statistics (1994)

36. Boulton, R.J., Gottliebsen, H., Hardy, R., Kelsey, T., Martin, U.: Design verification for control engineering. In: Boiten, E.A., Derrick, J., Smith, G. (eds.) Conference on Integrated Formal Methods. Lecture Notes in Computer Science, vol. 2999, pp. 21–35. Springer, Berlin (2004)

37. Bovier, A.: Metastability. Lecture Notes in Mathematics, vol. 1970. Springer, Berlin (2009)

38. Branicky, M.S.: Studies in hybrid systems: modeling, analysis and control. Sc.D. Thesis, MIT (June 1995)

39. Branicky, M.S., Borkar, V.S., Mitter, S.K.: A unified framework for hybrid control: model and optimal control theory. IEEE Trans. Autom. Control **43**(1), 31–45 (1998)

40. Bryant, R.E.: Graph-based algorithms for boolean function manipulation. IEEE Trans. Comput. **35**(8), 677–691 (1986)

41. Bujorianu, M.L.: Capacities and Markov processes. Libertas Math. **24**, 201–210 (2004)
42. Bujorianu, M.L., Lygeros, J.: Reachability questions in piecewise deterministic Markov processes. In: Maler, O., Pnueli, A. (eds.) Hybrid Systems: Computation and Control. Lecture Notes in Computer Science, vol. 2623, pp. 126–140. Springer, Berlin (2003)
43. Bujorianu, M.L.: Extended stochastic hybrid systems and their reachability problem. In: Alur, R., Pappas, G. (eds.) Hybrid Systems: Computation and Control. Lecture Notes in Computer Science, vol. 2993, pp. 234–249. Springer, Berlin (2004)
44. Bujorianu, M.L., Lygeros, J.: General stochastic hybrid systems: modelling and optimal control. In: Proc. of 43rd IEEE Conference in Decision and Control, pp. 182–187 (2004)
45. European Commission COLUMBUS project: Design of embedded controllers for safety critical systems. www.columbus.gr
46. Bujorianu, M.L., Lygeros, J.: New insights on stochastic reachability. In: Proc. of 46th IEEE Conference in Decision and Control, pp. 6172–6177 (2007)
47. Bujorianu, M.L.: Dealing with stochastic reachability. In: Proc. of 48th IEEE Conference in Decision and Control, pp. 2935–2940 (2009)
48. Chen, Z.Q., Ma, Z.-M., Rockner, M.: Quasi-homeomorphisms of Dirichlet forms. Nagoya Math. J. **136**, 1–15 (1994)
49. Chen, T., Han, T., Mereacre, A., Katoen, J.-P.: Model checking of continuous-time Markov chains against timed automata specifications. Log. Methods Comput. Sci. **7**(1), 1–34 (2011)
50. Cherkasov, I.: Transformation of diffusion equations by Kolmogorov's method. Sov. Math. Dokl. **21**(1), 175–179 (1980)
51. Choquet, G.: Theory of capacities. Ann. Inst. Fourier **5**, 131–291 (1953)
52. Chrisman, L.: Incremental conditioning of lower and upper probabilities. Int. J. Approx. Reason. **13**(1), 1–25 (1995)
53. Chung, K.L.: Probabilistic approach in potential theory to the equilibrium problem. Ann. Inst. Fourier **23**, 313–322 (1973)
54. Chung, K.L.: Green, Brown and Probability. World Scientific, Singapore (1995)
55. Chutinan, A., Krogh, B.H.: Verification of polyhedral-invariant hybrid automata using polygonal flow pipe approximations. In: Vaandrager, F.W., van Schuppen, J.H. (eds.) Hybrid Systems: Computation and Control. Lecture Notes in Computer Science, vol. 1569, pp. 76–90. Springer, Berlin (1999)
56. Clarke, E.M., Emerson, E.A., Sistla, A.P.: Automatic verification of finite-state concurrent systems using temporal logic specifications. ACM Trans. Program. Lang. Syst. **8**(2), 244–263 (1986)
57. Congdon, P.: Applied Bayesian Modelling. Wiley Series in Probability and Statistics (2003)
58. Costa, O.L.V., Dufour, F.: On the Poisson equation for piecewise-deterministic Markov processes. SIAM J. Control Optim. **42**(3), 985–1001 (2003)
59. Costa, O.L.V., Raymundo, C.A.B., Dufour, F.: Optimal stopping with continuous control of piecewise deterministic Markov processes. Stoch. Stoch. Rep. **70**(1–2), 41–73 (2000)
60. Costa, O.L.V., Davis, M.H.A.: Approximations for optimal stopping of a piecewise-deterministic process. Math. Control Signals Syst. **1**(2), 123–146 (1988)
61. Cozman, F.G.: Computing posterior upper expectations. Int. J. Approx. Reason. **24**, 191–205 (2000)
62. Dang, T., Maler, O.: Reachability analysis via face lifting. In: Sastry, S., Henzinger, T.A. (eds.) Hybrid Systems: Computation and Control. Lecture Notes in Computer Science, vol. 1386, pp. 96–109. Springer, Berlin (1998)
63. Davis, M.H.A., Vellekoop, M.H.: Permanent health insurance: a case study in piecewise-deterministic Markov modelling. Mitt. - Schweiz. Ver. Versicher.math. **2**, 177–212 (1995)
64. Davis, M.H.A.: Markov Models and Optimization. Chapman & Hall, London (1993)
65. Davis, M.H.A.: Piecewise-deterministic Markov processes: a general class of non-diffusion stochastic models. J. R. Stat. Soc. B, **46**(3), 353–388 (1984)
66. Davis, M.H.A., Dempster, V., Sethi, S.P., Vermes, D.: Optimal capacity expansion under uncertainty. Adv. Appl. Probab. **19**, 156–176 (1987)
67. Dellacherie, C., Maisonneuve, B., Meyer, P.-A.: Probabilités et Potentiel. Hermann, Paris (1992) (Ch. XVII à XXIV. Processus de Markov (fin). Complements de calcul stochastique)

68. Dempster, D.: Upper and lower probabilities induced by a multi-valued mapping. Ann. Math. Stat. **38**, 325–339 (1967)
69. Donsker, M.D., Varadhan, S.R.S.: Asymptotic evaluation of certain Markov process expectations for large time I. Commun. Pure Appl. Math. **28**, 1–47 (1975)
70. Donsker, M.D., Varadhan, S.R.S.: Asymptotic evaluation of certain Markov process expectations for large time II. Commun. Pure Appl. Math. **28**, 279–301 (1975)
71. Donsker, M.D., Varadhan, S.R.S.: Asymptotic evaluation of certain Markov process expectations for large time III. Commun. Pure Appl. Math. **29**, 389–461 (1976)
72. Donsker, M.D., Varadhan, S.R.S.: Asymptotic evaluation of certain Markov process expectations for large time IV. Commun. Pure Appl. Math. **36**, 183–212 (1983)
73. Doob, J.L.: Stochastic Processes. Wiley, New York (1953)
74. Doob, J.L.: Classical Potential Theory and Its Probabilistic Counterpart. Springer, Berlin (1984)
75. Durbin, J., Williams, D.: The first-passage density of the Brownian motion process to a curved boundary. J. Appl. Probab. **29**, 291–304 (1992)
76. Dynkin, E.B.: Optimal choice of the stopping moment of a Markov process. Dokl. Akad. Nauk SSSR **150**, 238–240 (1963)
77. Dynkin, E.B.: Markov Processes I. Springer, Berlin (1965)
78. Dynkin, E.B., Yushkevich, A.A.: Markov Processes: Theorems and Problems. Plenum, New York (1969)
79. El Karoui, N., Lepeltier, J.-P., Millet, A.: A probabilistic approach to the reduite in optimal stopping. Probab. Math. Stat. **13**(1), 97–121 (1992)
80. El-Samad, H., Fazel, M., Liu, X., Papachristodoulou, A., Prajna, S.: Stochastic reachability analysis in complex biological networks. In: Proc. of the IEEE American Control Conference (2006) (6 pp.)
81. El Kaabouchi, A.: Mesures dominée par une capacité alternée d'ordre 2. Proc. Am. Math. Soc. **121**(3), 823–832 (1994)
82. Ephraim, Y., Merhav, N.: Hidden Markov processes. IEEE Trans. Inf. Theory **48**(6), 1518–1569 (2002) (Special Issue on Shannon Theory: perspectives, trends, and applications)
83. Ethier, S.N., Kurtz, T.G.: Markov Processes: Characterization and Convergence. Wiley, New York (1986)
84. Everdij, M., Blom, H.A.P.: Bias and uncertainty in accident risk assessment. NLR Report CR-2002-137, National Laboratory of Aerospace NLR (2002)
85. Everdij, M., Blom, H.A.P.: Modelling hybrid state Markov processes through dynamically and stochastically coloured Petri nets. HYBRIDGE Deliverable D2.4 (2005). Available on [126]
86. Everdij, M., Blom, H.A.P., Stroeve, S.H.: Structured assessment of bias and uncertainty in Monte Carlo simulated ask risk. In: Proc. of the 8th Int. Conf. on Probabilistic Safety Assessment and Management (2006)
87. Farid, M., Davis, M.H.A.: Optimal consumption and exploration: a case study in piecewise-deterministic Markov modelling. In: Optimal Control and Differential Games, Vienna, 1997. Ann. Oper. Res., vol. 88, pp. 121–137 (1999)
88. Fine, T.L.: Lower probability models for uncertainty and non-deterministic processes. J. Stat. Plan. Inference **20**, 389–411 (1988)
89. Fitzsimmons, P.J.: Markov processes with equal capacities. J. Theor. Probab. **12**, 271–292 (1999)
90. Fleming, W., Soner, H.: Controlled Markov Processes and Viscosity Solutions. Springer, New York (1993)
91. Fränzle, M., Hermanns, H., Teige, T.: Stochastic satisfiability modulo theory: a novel technique for the analysis of probabilistic hybrid systems. In: Hybrid Systems: Computation and Control. Lecture Notes in Computer Science, pp. 172–186. Springer, Berlin (2008)
92. Fränzle, M., Herde, C.: Efficient proof engines for bounded model checking of hybrid systems. In: Workshop on Formal Methods for Industrial Critical Systems (2004). Electron. Notes Theor. Comput. Sci. **133**, 119–137 (2005)

93. Fukushima, M.: Dirichlet Forms and Markov Processes. North-Holland, Amsterdam (1980)
94. Gatarek, D.: On first order quasi-variational inequalities with integral terms. Appl. Math. Optim. **24**, 85–98 (1992)
95. Garroni, M.G., Menaldi, J.L.: Second Order Elliptic Integro-Differential Problems. Chapman & Hall/CRC Press, London/Boca Raton (2002)
96. Getoor, R.K.: Markov Processes: Ray Processes and Right Processes. Lecture Notes in Mathematics, vol. 440. Springer, Berlin (1975)
97. Getoor, R.K.: Excessive Measures. Birkhäuser, Boston (1990)
98. Getoor, R.K., Glover, J.: Riesz decompositions in Markov process theory. Trans. Am. Math. Soc. **285**(1), 107–132 (1984)
99. Getoor, R.K., Steffens, J.: The energy functional, balayage, and capacity. Ann. Henri Poincaré **23**(2), 321–357 (1987)
100. Getoor, R.K.: Excursions of a Markov process. Ann. Probab. **7**(2), 244–266 (1979)
101. Glover, J.: Representing last exit potentials as potentials of measures. Z. Wahrscheinlichkeitstheor. Verw. Geb. **61**(1), 17–30 (1982)
102. Graversen, S.E.: A Riesz decomposition theorem. Nagoya Math. J. **114**, 123–133 (1989)
103. Ghosh, M.K., Arapostathis, A., Marcus, S.I.: An optimal control problem arising in flexible manufacturing systems. In: Proc. of 30th IEEE Conf. on Decision and Control, pp. 1844–1849 (1991)
104. Ghosh, M.K., Arapostathis, A., Marcus, S.I.: Optimal control of switching diffusions with application to flexible manufacturing systems. SIAM J. Control Optim. **31**(5), 1183–1204 (1993)
105. Ghosh, M.K., Arapostathis, A., Marcus, S.I.: Ergodic control of switching diffusions. SIAM J. Control Optim. **35**(6), 1952–1988 (1997)
106. Gihman, I.I.: On the theory of differential equations of random processes. Ukr. Math. J. **2**(4), 37–63 (1950)
107. Gokbayrak, K., Cassandras, C.G.: Stochastic optimal control of a hybrid manufacturing system model. In: Proc. of the 38th Conference on Decision and Control, pp. 919–924 (1999)
108. Grecea, V.: On some results concerning the reduite and balayage. Math. Nachr. **223**, 66–75 (2001)
109. Grimmett, G., Stirzaker, D.: Probability and Random Processes. Oxford University Press, London (1982)
110. Gugerli, U.S.: Optimal stopping of a piecewise-deterministic Markov process. Stochastics **19**(4), 221–236 (1986)
111. Havelund, K., Larsen, K.G., Skou, A.: Formal verification of a power controller using the real-time model checker UPAAL. In: Katoen, J.P. (ed.) Formal Methods for Real-Time and Probabilistic Systems. Lecture Notes in Computer Science, vol. 1601, pp. 277–298. Springer, Berlin (1999)
112. Hedlund, S., Rantzer, A.: Convex dynamic programming for hybrid systems. IEEE Trans. Autom. Control **47**(9), 1536–1540 (2002)
113. Helmes, K., Röhl, S., Stockbridge, R.H.: Computing moments of the exit time distribution for Markov processes by linear programming. Math. Methods Oper. Res. **49**(4), 516–530 (2001)
114. Henzinger, T.A., Ho, P.-H., Wong-Toi, H.: A user guide to HYTECH. In: Brinksma, E., Cleaveland, R., Larsen, K.G., Maragaria, T., Steffen, B. (eds.) TACAS 95: Tools and Algorithms for the Construction and Analysis of Systems. Lecture Notes in Computer Science, vol. 1019, pp. 41–71. Springer, Berlin (1995)
115. Henzinger, T.A.: The theory of hybrid automata. In: Proceedings 11th Logic in Computer Science, pp. 278–292. IEEE Comput. Soc., Los Alamitos (1996)
116. Henzinger, T.A., Kopke, P.W., Puri, A., Varaiya, P.: What's decidable about hybrid automata? J. Comput. Syst. Sci. **57**(1), 94–124 (1998)
117. Hespanha, J.P.: Stochastic hybrid systems: application to communication networks. In: Alur, R., Pappas, G. (eds.) Hybrid Systems: Computation and Control. Lecture Notes in Computer Science, vol. 2993, pp. 387–401. Springer, Berlin (2004)

118. Hofbaur, M.W., Williams, B.C.: Mode estimation of probabilistic hybrid systems. In: Tomlin, C.J., Greenstreet, M.R. (eds.) Hybrid Systems: Computation and Control. Lecture Notes in Computer Science, vol. 2289, pp. 253–266. Springer, Berlin (2002)

119. Hoekstra, J.M., Ruigrok, R.C.J., van Gent, R.N.H.W.: Free flight in a crowded airspace? In: 3rd USA/Europe Air Traffic Management R&D Seminar Napoli (2000)

120. Huber, P.J., Strassen, V.: Minimax tests and the Neyman–Pearson lemma for capacities. Ann. Stat. **1**, 251–263 (1973)

121. Huber, P.J.: Robust Statistics. Wiley, New York (1980)

122. Hu, J., Lygeros, J., Sastry, S.: Towards a theory of stochastic hybrid systems. In: Lynch, N., Krogh, B. (eds.) Hybrid Systems: Computation and Control. Lecture Notes in Computer Science, vol. 1790, pp. 160–173. Springer, Berlin (2000)

123. Hu, J., Wu, W., Sastry, S.: Modeling subtilin production in *Bacillus subtilis* using stochastic hybrid systems. In: Alur, R., Pappas, G. (eds.) Hybrid Systems: Computation and Control. Lecture Notes in Computer Science, vol. 2993, pp. 163–166. Springer, Berlin (2004)

124. Hu, J., Prandini, M., Sastry, S.: Probabilistic safety analysis in three dimensional aircraft flight. In: Proc. of the 42nd IEEE Conference on Decision and Control, pp. 5335–5340 (2003)

125. Hu, J., Prandini, M., Sastry, S.: Aircraft conflict prediction in the presence of a spatially correlated wind field. IEEE Trans. Intell. Transp. Syst. **6**(3), 326–340 (2005)

126. European Commission HYBRIDGE project: Distributed control and stochastic analysis of hybrid systems supporting safety critical real-time system design. www.nlr.nl/public/hosted-sites/hybridge/

127. Hu, J., Prandini, M.: Aircraft conflict detection: a method for computing the probability of conflict based on Markov chain approximation. In: Proc. of the European Control Conference (2003)

128. Hunt, G.A.: Markov processes and potentials I. Ill. J. Math. **1**(1), 44–93 (1957)

129. Hunt, G.A.: Markov processes and potentials II. Ill. J. Math. **1**(3), 316–369 (1957)

130. Hunt, G.A.: Markov processes and potentials III. Ill. J. Math. **2**(2), 151–213 (1958)

131. Huth, M., Ryan, M.: Logic in Computer Science: Modelling and Reasoning about Systems. Cambridge University Press, Cambridge (2000)

132. Hwang, I., Hwang, J., Tomlin, C.J.: Flight-model-based aircraft conflict detection using a residual-mean interacting multiple model algorithm. In: AIAA Guidance, Navigation, and Control Conference AIAA-2003-5340 (2003)

133. European Commission iFLY project: Safety, complexity and responsibility based design and validation of highly automated air traffic management. http://ifly.nlr.nl/

134. Ikeda, N., Nagasawa, M., Watanabe, S.: Construction of Markov processes by piecing out. Proc. Jpn. Acad. **42**, 370–375 (1966)

135. Iscoe, I., McDonald, D.: Induced Dirichlet forms and capacitary inequalities. Ann. Probab. **18**(3), 1195–1221 (1990)

136. Itô, K.: Differential equations determining a Markov process. J. Pan-Jpn. Math. Coll. **1077** (1942) in Japanese; in English Kiyosi Ito Selected Papers. Springer, Berlin (1986)

137. Jacod, J., Shiryayev, A.N.: Limit Theorems for Stochastic Processes. Springer, New York (1987)

138. Julius, A.A., Pappas, G.J.: Approximate abstraction of stochastic hybrid systems. IEEE Trans. Autom. Control **54**(6), 1193–1203 (2009)

139. Julius, A.A., D'Innocenzo, A., Di Benedetto, M.D., Pappas, G.J.: Approximate equivalence and synchronization of metric transition systems. Syst. Control Lett. **58**, 94–101 (2009)

140. Julius, A.A.: Approximate abstraction of stochastic hybrid automata. In: Hespanha, J.P., Tiwari, A. (eds.) Hybrid Systems: Computation and Control. Lecture Notes in Computer Science, vol. 3927, pp. 318–332. Springer, Berlin (2006)

141. Khinchine, A.Y.: Asymptotische Gesetze der Wahrscheinlichkeitsrechnung. Ergeb. Math. **77**(4) (1933)

142. Kinderlehrer, D., Stampacchia, G.: An Introduction to Variational Inequalities and Their Applications. Academic Press, London (1980)

143. Kloeden, P., Platen, E.: Numerical Solution of Stochastic Differential Equations. Springer, Berlin (1992)

144. Kolmogoroff, A.: Über die analytischen Methoden in der Wahrscheinlichkeitsrechnung. Math. Anal. **104**, 415–458 (1931)

145. Koutsoukos, X.D.: Optimal control of stochastic hybrid systems based on locally consistent Markov decision processes. In: Proc. of 43rd IEEE Conference in Decision and Control (2004)

146. Koutsoukos, X., Riley, D.: Computational methods for verification of stochastic hybrid systems. IEEE Trans. Syst. Man Cybern., Part A, Syst. Hum. **38**(2), 385–396 (2008)

147. Kruzynski, M., Chelminiak, P.: Mean first-passage time in the stochastic theory of biochemical processes. J. Stat. Phys. **110**, 137–181 (2003)

148. Krystul, J., Blom, H.A.P.: Sequential Monte Carlo simulation of rare event probability in stochastic hybrid systems. Preprint, 16th IFAC World Congress (2005)

149. Krystul, J., Bagchi, A.: Approximation of first passage times of switching diffusion. In: Proc. 16th International Symposium on Mathematical Theory of Networks and Systems (2004)

150. Kwiatkowska, M., Norman, G., Parker, D.: Probabilistic symbolic model checking with PRISM: a hybrid approach. International. J. on Software Tools Technology Transfer **6**, 128–142 (2004)

151. Kwiatkowska, M., Norman, G., Segala, R., Sproston, J.: Verifying quantitative properties of continuous probabilistic timed automata. In: Palamidessi, C. (ed.) International Conference on Concurrency Theory. Lecture Notes in Computer Science, vol. 1877, pp. 123–137. Springer, Berlin (2000)

152. Kwiatkowska, M., Norman, G., Segala, R., Sproston, J.: Automatic verification of real-time systems with discrete probability distributions. Theor. Comput. Sci. **282**, 101–150 (2002)

153. Kunita, H., Watanabe, T.: Markov processes and Martin boundaries. Part I. Ill. J. Math. **9**, 485–526 (1965)

154. Kunita, H., Watanabe, T.: Notes on transformations of Markov processes connected with multiplicative functionals. Mem. Fac. Sci., Kyushu Univ., Ser. A, Math. **17**(2), 181–191 (1963)

155. Kurtz, T.G.: Equivalence of stochastic equations and martingale problems. In: Crisan, D. (ed.) Stochastic Analysis 2010, pp. 113–130. Springer, Berlin (2011)

156. Kurzhanski, A.B., Varaiya, P.: Ellipsoidal techniques for reachability analysis. In: Lynch, N., Krogh, B. (eds.) Hybrid Systems: Computation and Control. Lecture Notes in Computer Science, vol. 1790, pp. 202–214. Springer, Berlin (2000)

157. Kurzhanski, A.B., Varaiya, P.: Ellipsoidal techniques for hybrid dynamics: the reachability problem. In: Dayawansa, W.P., Lindquist, A., Zhou, Y. (eds.) New Directions and Applications in Control Theory. Lecture Notes in Control and Information Sciences, vol. 321, pp. 193–205. Springer, Berlin (2005)

158. Kushner, H.J.: Probability Methods for Approximations in Stochastic Control and for Elliptic Equations. Academic Press, New York (1977)

159. Kyburg, H.E. Jr.: Bayesian and non-Bayesian evidential updating. Artif. Intell. **31**, 271–293 (1987)

160. Ishii, H.: Perron's method for Hamilton–Jacobi equations. Duke Math. J. **55**(2), 369–384 (1987)

161. Lafferriere, G., Pappas, G.J., Sastry, S.: Reachability analysis of hybrid systems using bisimulations. In: Proc. of the IEEE Conference on Decision and Control, pp. 1623–1628 (1998)

162. Lavine, M.: Sensitivity in Bayesian statistics the prior and the likelihood. J. Am. Stat. Assoc. **86**(414), 396–399 (1991)

163. Lemmon, M., Stiver, J.A., Antsaklis, P.J.: Event identification and intelligent hybrid control. In: Grossman, R.L., Nerode, A., Ravn, A.P., Rischel, H. (eds.) Hybrid Systems. Lecture Notes in Computer Science, vol. 736, pp. 268–296. Springer, Berlin (1993)

164. Lenhard, S.M., Yamada, N.: Perron's method for viscosity solutions associated with piecewise deterministic processes. Funkc. Ekvacioj **34**, 173–186 (1991)

165. Lenhart, S., Liao, Y.C.: Integro-differential equations associated with optimal stopping time of a piecewise-deterministic process. Stochastics **15**(3), 183–207 (1985)

166. Levi, I.: The Enterprise of Knowledge. MIT Press, Cambridge (1980)
167. Limnios, N., Oprisan, G.: Semi-Markov Processes and Reliability. Birkhäuser, Boston (2001)
168. Lygeros, J., Johansson, K.H., Simić, S.N., Zhang, J., Sastry, S.: Dynamical properties of hybrid automata. IEEE Trans. Autom. Control **48**, 2–17 (2003)
169. Lygeros, J.: Lecture notes on hybrid systems (2004)
170. Lygeros, J., Tomlin, C., Sastry, S.: Controllers for reachability specifications of hybrid systems. Automatica **35**, 349–370 (1999)
171. Lygeros, J., Tomlin, C., Sastry, S.: A game theoretic approach to controller design for hybrid systems. Proc. IEEE **88**, 949–969 (2000)
172. Lymperopoulos, I., Lygeros, J.: Sequential Monte Carlo methods for multi-aircraft trajectory prediction in air traffic management. Int. J. Adapt. Control Signal Process. **24**, 830–849 (2010)
173. Lymperopoulos, I., Lygeros, J.: Improved multi-aircraft ground trajectory prediction for air traffic control. J. Guid. Control Dyn. **33**, 347–362 (2010)
174. Lymperopoulos, I., Lygeros, J.: Improved ground trajectory prediction by multi-aircraft track fusion for air traffic control. In: AIAA Guidance, Navigation and Control Conference and Exhibit AIAA-2009-5784 (2009)
175. Lymperopoulos, I., Lygeros, J.: Adaptive aircraft trajectory prediction using particle filters. In: AIAA Guidance, Navigation and Control Conference and Exhibit (2009)
176. Ma, M., Rockner, M.: The Theory of (Non-Symmetric) Dirichlet Forms and Markov Processes. Springer, Berlin (1990)
177. Manna, Z., Pnueli, A.: The temporal Logic of Reactive and Concurrent Systems Specification. Springer, New York (1992)
178. Malhame, R., Chong, C.-Y.: Electric load model synthesis by diffusion approximation of a high-order hybrid-state stochastic system. IEEE Trans. Autom. Control **9**(9), 854–860 (1985)
179. Mataloni, S.: Representation formulas for non-symmetric Dirichlet forms. Z. Anal. Anwend. **18**(4), 1039–1064 (1999)
180. Mertens, J.F.: Strongly supermedian functions and optimal stopping. Z. Wahrscheinlichkeitstheor. Verw. Geb. **22**, 45–68 (1972)
181. Meyer, P.A.: Probability and Potentials. Blaisdell, Waltham (1966)
182. Meyer, P.A.: Renaissance, recollectments, melanges, ralentissement de processus de Markov. Ann. Inst. Fourier **25**, 465–497 (1975)
183. Meyer, P.A.: Processus de Markov. Lecture Notes in Mathematics, vol. 26. Springer, Berlin (1967)
184. Mitchell, I., Tomlin, C.: Level set methods for computation in hybrid systems. In: Lynch, N., Krogh, B. (eds.) Hybrid Systems: Computation and Control. Lecture Notes in Computer Science, vol. 1790, pp. 310–323. Springer, Berlin (2000)
185. Mordecki, E., Salminen, P.: Optimal stopping of Hunt and Lévy processes. Stochastics **79**(3–4), 233–251 (2007)
186. Nagai, H.: On an optimal stopping problem and a variational inequalities. J. Math. Soc. Jpn. **30**(2), 303–312 (1978)
187. Nardo, E.D., Nobile, A.G., Pirozzi, E., Ricciardi, L.M.: A computational approach to first passage time problems for Gauss–Markov processes. Adv. Appl. Probab. **33**, 453–482 (2001)
188. Nerode, A., Kohn, W.: Models for hybrid systems: automata, topologies, controllability, observability. In: Grossman, R.L., Nerode, A., Ravn, A.P., Rischel, H. (eds.) Hybrid Systems: Computation and Control. Lecture Notes in Computer Science, vol. 736, pp. 317–356. Springer, Berlin (1993)
189. Nobile, A.G., Ricciardi, L.M., Sacerdote, L.: Exponential trends of Ornstein–Uhlenbeck first-passage-time densities. J. Appl. Probab. **22**, 360–369 (1985)
190. Peskir, G., Shiryayev, A.N.: Optimal Stopping and Free-Boundary Problems. Birkhäuser, Basel (2006)

191. Petrowsky, I.: Über das Irrfahrtproblem. Math. Ann. **109**, 425–444 (1934)
192. Perron, O.: Eine neue Behandlung der ersten Randwerteaufgabe für $\Delta u = 0$. Math. Z. **18**, 42–54 (1923)
193. Pola, G., Bujorianu, M.L., Lygeros, J., Di Benedetto, M.D.: Stochastic hybrid models: an overview with applications to air traffic management. In: Proc. of the Analysis and Design of Hybrid Systems, pp. 45–50. IFAC Elsevier, Amsterdam (2003)
194. Prajna, S., Jadbabaie, A., Pappas, G.J.: Stochastic safety verification using barrier certificates. In: Proc. of the IEEE Conference on Decision and Control, pp. 929–934 (2004)
195. Prajna, S., Papachristodoulou, A., Parrilo, P.A.: Introducing SOSTOOLS: a general purpose sum of squares programming solver. In: Proc. of the IEEE Conference on Decision and Control, pp. 741–746 (2002)
196. Prandini, M., Hu, J.: A Stochastic approximation method for reachability computations. In: [28], pp. 107–139
197. Prandini, M., Hu, J.: Applications of reachability analysis for stochastic hybrid systems to aircraft conflict prediction. IEEE Trans. Autom. Control **54**(4), 913–917 (2009)
198. Puri, A., Varaiya, P.: Verification of hybrid systems using abstractions. In: Antsaklis, P.J. (ed.) Hybrid Systems II. Lecture Notes in Computer Science, vol. 999, pp. 359–369. Springer, Berlin (1995)
199. Ricciardi, L.: On the transformation of diffusion processes into the Wiener process. J. Math. Anal. Appl. **54**, 185–189 (1976)
200. Rogers, L.C.G., Pitman, J.W.: Markov functions. Ann. Probab. **9**(4), 573–582 (1981)
201. Riley, D.: Modeling, simulation, and verification of biochemical processes using stochastic hybrid systems. PhD Dissertation (2009)
202. Riley, D., Koutsoukos, X., Riley, K.: Reachability analysis of a biodiesel production system using stochastic hybrid systems. In: Proc. of 15th IEEE Mediterranean Conference on Control and Automation (2007)
203. Riley, D., Koutsoukos, X., Riley, K.: Safety analysis of sugar cataract developement using stochastic hybrid systems. In: Bemporad, A., Bicchi, A., Buttazzo, G.C. (eds.) Hybrid Systems: Computation and Control. Lecture Notes in Computer Science, vol. 4416, pp. 758–761. Springer, Berlin (2007)
204. Riley, D., Koutsoukos, X., Riley, K.: Reachability analysis for stochastic hybrid systems using multilevel splitting. In: Majumdar, R., Tabuada, P. (eds.) Hybrid Systems: Computation and Control, vol. 5469, pp. 460–464 (2009)
205. Riley, D., Koutsoukos, X.: Simulation of stochastic hybrid systems using probabilistic boundary detection and adaptive time stepping. Simul. Model. Pract. Theory **18**(9), 1397–1411 (2010)
206. Savage, L.J.: The Foundations of Statistics. Dover, New York (1972)
207. Salminen, P.: Optimal stopping of one-dimensional diffusions. Math. Nachr. **124**, 85–101 (1985)
208. Sharpe, M.: General Theory of Markov Processes. Academic Press, San Diego (1988)
209. Schmeidler, D.: Subjective probability and expected utility without additivity. Econometrica **57**, 571–587 (1989)
210. Seidenfeld, T., Schervish, M., Kadane, J.B.: A representation of partially ordered preferences. Ann. Stat. **23**(6), 2168–2217 (1995)
211. Shafer, G.: A Mathematical Theory of Evidence. Princeton University Press, Princeton (1976)
212. Shiryayev, A.N.: Optimal Stopping Rules. Springer, Berlin (1976)
213. Shtrichman, O.: Tuning SAT checkers for bounded model checking. In: Emerson, E.A., Sistla, A.P. (eds.) Computer Aided Verification. Lecture Notes in Computer Science, vol. 1855, pp. 480–494. Springer, Berlin (2000)
214. Siegrist, K.: Random evolution processes with feedback. Trans. Am. Math. Soc. **265**(2), 375–392 (1981)
215. Sinai, Ya.G.: Dynamical Systems, Ergodic Theory and Applications. Springer, Berlin (2000)
216. Sproston, J.: Analysing subclasses of probabilistic hybrid automata. In: Kwiatkowska, M.Z. (ed.) 2nd International Workshop on Probabilistic Methods in Verification (1999)

217. Stroock, D.W., Varadhan, S.R.S.: Multidimensional Diffusion Processes. Springer, Berlin (1979)
218. Suppes, P.: The measurement of belief. J. R. Stat. Soc. B **2**, 160–191 (1974)
219. Syski, R.: Passage Times for Markov Chains. Studies in Probability, Optimization and Statistics, vol. 1 (1991) (556 pp.)
220. Tabuada, P., Pappas, G.J., Lima, P.: Compositional abstractions of hybrid control systems. Discrete Event Dyn. Syst. Theory Appl. **14**, 203–238 (2004)
221. Tabuada, P., Ames, A., Julius, A.A., Pappas, G.J.: Approximate reduction of dynamical systems. Syst. Control Lett. **7**(57), 538–545 (2008)
222. Taira, K.: Boundary value problems for elliptic integro-differential operator. Math. Z. **222**, 305–327 (1996)
223. Varaiya, P.: Reach set computation using optimal control problems in multidimensional systems. In: Proc. of KIT Workshop on Verification of Hybrid Systems (1998)
224. Zhou, K., Doyle, J.C., Glover, K.: Robust and Optimal Control. Prentice Hall, Upper Saddle River (1996)
225. Wasserman, L.A., Kadane, J.B.: Bayes's theorem for Choquet capacities. Ann. Stat. **18**(3), 1328–1339 (1990)
226. Watkins, O.J.: Stochastic reachability, conflict detection, and air traffic management. Doctoral thesis (2004)
227. Walley, P.: Statistical Reasoning with Imprecise Probabilities. Chapman & Hall, London (1991)
228. Weis, L., Werner, D.: Reference measures and the fine topology. Preprint (1999)

Index

L.M. Bujorianu, *Stochastic Reachability Analysis of Hybrid Systems*,
Communications and Control Engineering,
DOI 10.1007/978-1-4471-2795-6, © Springer-Verlag London Limited 2012